Ben Stacy Jerrik (Ed.)

126749 Johnjones

AF694320

Ben Stacy Jerrik (Ed.)

# 126749 Johnjones

## Small Solar System Body, Centaur (Minor Planet), Kuiper Belt

Part Press

**Imprint**

Permission is granted to copy, distribute and/or modify this document under the terms of the GNU Free Documentation License, Version 1.2 or any later version published by the Free Software Foundation; with no Invariant Sections, with the Front-Cover Texts, and with the Back- Cover Texts. A copy of the license is included in the section entitled "GNU Free Documentation License".

All parts of this book are extracted from Wikipedia, the free encyclopedia (www.wikipedia.org).

You can get detailed informations about the authors of this collection of articles at the end of this book. The editors (Ed.) of this book are no authors. They have not modified or extended the original texts.

Pictures published in this book can be under different licences than the GNU Free Documentation License. You can get detailed informations about the authors and licences of pictures at the end of this book.

The content of this book was generated collaboratively by volunteers. Please be advised that nothing found here has necessarily been reviewed by people with the expertise required to provide you with complete, accurate or reliable information. Some information in this book maybe misleading or wrong. The Publisher does not guarantee the validity of the information found here. If you need specific advice (f.e. in fields of medical, legal, financial, or risk management questions) please contact a professional who is licensed or knowledgeable in that area.

Any brand names and product names mentioned in this book are subject to trademark, brand or patent protection and are trademarks or registered trademarks of their respective holders. The use of brand names, product names, common names, trade names, product descriptions etc. even without a particular marking in this works is in no way to be construed to mean that such names may be regarded as unrestricted in respect of trademark and brand protection legislation and could thus be used by anyone.

Cover image: www.ingimage.com
Concerning the licence of the cover image please contact ingimage.

Publisher:
Part Press is a trademark of
International Book Market Service Ltd., 17 Rue Meldrum, Beau Bassin, 1713-01 Mauritius
Email: info@bookmarketservice.com
Website: www.bookmarketservice.com

Published in 2013

Printed in: U.S.A., U.K., Germany. This book was not produced in Mauritius.

**ISBN: 978-613-9-09189-8**

# Contents

## Articles



## References

# 12446_Juliabryant

| Discovery and designation | |
|---|---|
| Discovered by | R. H. McNaught and J. B. Child |
| Discovery site | Macquarie |
| Discovery date | August 15, 1996 |
| **Designations** | |
| MPC designation | 12446 |
| Pronunciation | I |
| Alternative names | 1996 PZ6 |
| **Orbital characteristics** | |
| Epoch May 14, 2008 | |
| Ap | 2.1737746 |
| Peri | 1.7200086 |
| Eccentricity | 0.1165360 |
| Orbital period | 992.2271406 |
| Mean anomaly | 98.30511 |
| Inclination | 23.81550 |
| Longitude of ascending node | 86.71410 |
| Argument of peri | 266.39027 |
| **Physical characteristics** | |
| Absolute magnitude *(H)* | 15.0 |

**12446 Juliabryant** (1996 PZ6) is an inner main-belt asteroid discovered on August 15, 1996 by R. H. McNaught and J. B. Child at Macquarie.

## References

## External links

- JPL Small-Body Database Browser on 12446 Juliabryant (http://ssd.jpl.nasa.gov/sbdb.cgi?sstr=12446+Juliabryant)

# Asteroid_belt

The **asteroid belt** is the region of the Solar System located roughly between the orbits of the planets Mars and Jupiter. It is occupied by numerous irregularly shaped bodies called asteroids or minor planets. The asteroid belt is also termed the **main asteroid belt** or **main belt** to distinguish its members from other asteroids in the Solar System such as near-Earth asteroids and trojan asteroids. About half the mass of the belt is contained in the four largest asteroids, Ceres, Vesta, Pallas, and Hygiea. These have mean diameters of more than 400 km, while Ceres, the asteroid belt's only dwarf planet, is about 950 km in diameter.[1] [2] [3] [4] The remaining bodies range down to the size of a dust particle. The asteroid material is so thinly distributed that numerous unmanned spacecraft have traversed it without incident. Nonetheless, collisions between large asteroids do occur, and these can form an asteroid family whose members have similar orbital characteristics and compositions. Nesvorny and Jenniskens (2010 Astrophysical Journal) attributed 85 percent of the Zodiacal Light dust to fragmentations of Jupiter-family comets, rather than from comets and collisions between asteroids in the asteroid belt. It was once thought that collisions of asteroids produce a fine dust that forms a major component of the zodiacal light. Individual asteroids within the asteroid belt are categorized by their spectra, with most falling into three basic groups: carbonaceous (C-type), silicate (S-type), and metal-rich (M-type).

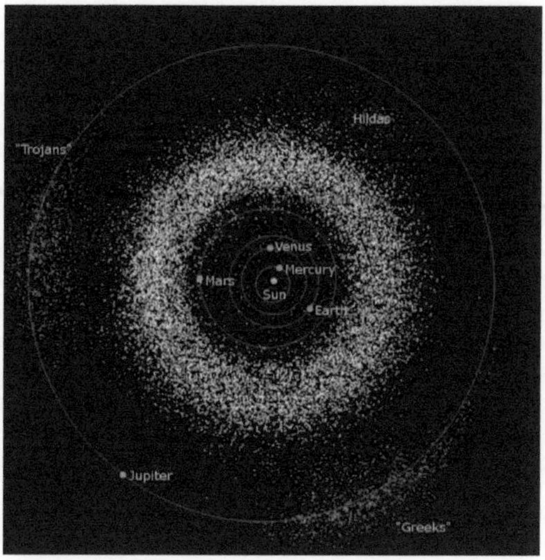

The asteroid belt (shown in white) is located between the orbits of Mars and Jupiter.

The asteroid belt formed from the primordial solar nebula as a group of planetesimals, the smaller precursors of the planets, which in turn formed protoplanets. Between Mars and Jupiter, however, gravitational perturbations from the giant planet imbued the protoplanets with too much orbital energy for them to accrete into a planet. Collisions became too violent, and instead of fusing together, the planetesimals and most of the protoplanets shattered. As a result, most of the asteroid belt's mass has been lost since the formation of the Solar System. Some fragments can eventually find their way into the inner Solar System, leading to meteorite impacts with the inner planets. Asteroid orbits continue to be appreciably perturbed whenever their period of revolution about the Sun forms an orbital resonance with Jupiter. At these orbital distances, a Kirkwood gap occurs as they are swept into other orbits.

Classes of small Solar System bodies in other regions include the centaurs, Kuiper belt and scattered disk objects, and Oort cloud comets.

## History of observation

In an anonymous footnote to his 1766 translation of Charles Bonnet's *Contemplation de la Nature*,[5] the astronomer Johann Daniel Titius of Wittenberg[6] [7] noted an apparent pattern in the layout of the planets. If one began a numerical sequence at 0, then included 3, 6, 12, 24, 48, etc., doubling each time, and added four to each number and divided by 10, this produced a remarkably close approximation to the radii of the orbits of the known planets as measured in astronomical units. This pattern, now known as the Titius–Bode law, predicted the semi-major axes of the six planets of the time (Mercury, Venus, Earth, Mars, Jupiter and Saturn) provided one allowed for a "gap" between the orbits of Mars and Jupiter. In his footnote Titius declared, "But should the Lord Architect have left that space empty? Not at all."[6] In 1768, the astronomer Johann Elert Bode made note of Titius's relationship in his *Anleitung zur Kenntniss des gestirnten Himmels* (English: *Instruction for the Knowledge of the Starry Heavens*) but did not credit Titius until later editions. It became known as "Bode's law".[7] When William Herschel discovered Uranus in 1781, the planet's orbit matched the law almost perfectly, leading astronomers to conclude that there had to be a planet between the orbits of Mars and Jupiter.

Giuseppe Piazzi, discoverer of Ceres, known as a planet for many years, then as asteroid number 1, and eventually, a dwarf planet

In 1800 the astronomer Baron Franz Xaver von Zach recruited 24 of his fellows into a club, the Vereinigte Astronomische Gesellschaft ("United Astronomical Society") which he informally dubbed the "Lilienthal Society"[8] for its meetings in Lilienthal, a small city near Bremen. Determined to bring the Solar System to order, the group became known as the "Himmelspolizei", or Celestial Police. Notable members included Herschel, the British Astronomer Royal Nevil Maskelyne, Charles Messier, and Heinrich Olbers.[9] The Society assigned to each astronomer a 15° region of the zodiac to search for the missing planet.[10]

Only a few months later, a non-member of the Celestial Police confirmed their expectations. On January 1, 1801, Giuseppe Piazzi, Chair of Astronomy at the University of Palermo, Sicily, found a tiny moving object in an orbit with exactly the radius predicted by the Titius–Bode law. He dubbed it Ceres, after the Roman goddess of the harvest and patron of Sicily. Piazzi initially believed it a comet, but its lack of a coma suggested it was a planet.[9] Fifteen months later, Heinrich Wilhelm Olbers discovered a second object in the same region, Pallas. Unlike the other known planets, the objects remained points of light even under the highest telescope magnifications, rather than resolving into discs. Apart from their rapid movement, they appeared indistinguishable from stars. Accordingly, in 1802 William Herschel suggested they be placed into a separate category, named *asteroids*, after the Greek *asteroeides*, meaning "star-like".[11] [12] Upon completing a series of observations of Ceres and Pallas, he concluded,[13]

> Neither the appellation of planets, nor that of comets, can with any propriety of language be given to these two stars ... They resemble small stars so much as hardly to be distinguished from them. From this, their asteroidal appearance, if I take my name, and call them Asteroids; reserving for myself however the liberty of changing that name, if another, more expressive of their nature, should occur.

Despite Herschel's coinage, for several decades it remained common practice to refer to these objects as planets.[5] By 1807, further investigation revealed two new objects in the region: 3 Juno and 4 Vesta.[14] The burning of Lilienthal in the Napoleonic wars brought this first period of discovery to a close,[14] and only in 1845 did astronomers detect another object (5 Astraea). Shortly thereafter new objects were found at an accelerating rate, and counting them among the planets became increasingly cumbersome. Eventually, they were dropped from the planet list as first suggested by Alexander von Humboldt in the early 1850s, and William Herschel's choice of nomenclature, "asteroids", gradually came into common use.[5]

The discovery of Neptune in 1846 led to the discrediting of the Titius–Bode law in the eyes of scientists, as its orbit was nowhere near the predicted position. To date, there is no scientific explanation for the law, and astronomers' consensus regards it as a coincidence.[15]

The expression "asteroid belt" came into use in the very early 1850s, although it is hard to pinpoint who coined the term. The first English use seems to be in the 1850 translation (by E. C. Otté) of Alexander von Humboldt's *Cosmos*:[16] "[...] and the regular appearance, about the 13th of November and the 11th of August, of shooting stars, which probably form part of a belt of asteroids intersecting the Earth's orbit and moving with planetary velocity". Other early appearances occur in Robert James Mann's *A Guide to the Knowledge of the Heavens*,[17] "The orbits of the asteroids are placed in a wide belt of space, extending between the extremes of [...]". The American astronomer Benjamin Peirce seems to have adopted that terminology and to have been one of its promoters.[18] One hundred asteroids had been located by mid-1868, and in 1891 the introduction of astrophotography by Max Wolf accelerated the rate of discovery still further.[19] A total of 1,000 asteroids had been found by 1921,[20] 10,000 by 1981,[21] and 100,000 by 2000.[22] Modern asteroid survey systems now use automated means to locate new minor planets in ever-increasing quantities.

# Origin

## Formation

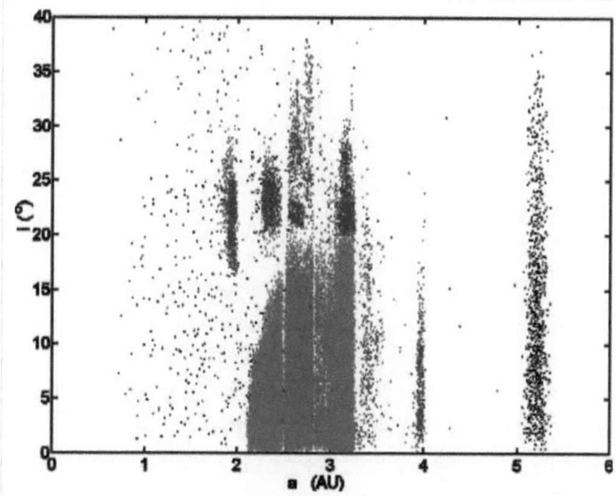


The asteroid belt showing the orbital inclinations versus distances from the Sun, with asteroids in the core region of the asteroid belt in red and other asteroids in blue

In 1802, shortly after discovering Pallas, Heinrich Olbers suggested to William Herschel that Ceres and Pallas were fragments of a much larger planet that once occupied the Mars–Jupiter region, this planet having suffered an internal explosion or a cometary impact many million years before.[23] Over time, however, this hypothesis has fallen from favor. The large amount of energy that would have been required to destroy a planet, combined with the belt's low combined mass, which is only about 4% of the mass of the Earth's Moon,[1] do not support the hypothesis. Further, the significant chemical differences between the asteroids are difficult to explain if they come from the same planet.[24] Today, most scientists accept that, rather than fragmenting from a progenitor planet, the asteroids never formed a planet at all.

In general in the Solar System, planetary formation is thought to have occurred via a process comparable to the long-standing nebular hypothesis: a cloud of interstellar dust and gas collapsed under the influence of gravity to form a rotating disk of material that then further condensed to form the Sun and planets.[25] During the first few million years of the Solar System's history, an accretion process of sticky collisions caused the clumping of small particles, which gradually increased in size. Once the clumps reached sufficient mass, they could draw in other bodies through gravitational attraction and become planetesimals. This gravitational accretion led to the formation of the rocky planets and the gas giants.

Planetesimals within the region which would become the asteroid belt were too strongly perturbed by Jupiter's gravity to form a planet. Instead they continued to orbit the Sun as before, while occasionally colliding.[26] In regions where the average velocity of the collisions was too high, the shattering of planetesimals tended to dominate over accretion,[27] preventing the formation of planet-sized bodies. Orbital resonances occurred where the orbital period of an object in the belt formed an integer fraction of the orbital period of Jupiter, perturbing the object into a different orbit; the region lying between the orbits of Mars and Jupiter contains many such orbital resonances. As Jupiter migrated inward following its formation, these resonances would have swept across the asteroid belt, dynamically exciting the region's population and increasing their velocities relative to each other.[28]

During the early history of the Solar System, the asteroids melted to some degree, allowing elements within them to be partially or completely differentiated by mass. Some of the progenitor bodies may even have undergone periods of explosive volcanism and formed magma oceans. However, because of the relatively small size of the bodies, the period of melting was necessarily brief (compared to the much larger planets), and had generally ended about 4.5 billion years ago, in the first tens of millions of years of formation.[29] In August 2007, a study of zircon crystals in an Antarctic meteorite believed to have originated from 4 Vesta suggested that it, and by extension the rest of the asteroid belt, had formed rather quickly, within ten million years of the Solar System's origin.[30]

## Evolution

The asteroids are not samples of the primordial Solar System. They have undergone considerable evolution since their formation, including internal heating (in the first few tens of millions of years), surface melting from impacts, space weathering from radiation, and bombardment by micrometeorites.[31] While some scientists refer to the asteroids as residual planetesimals,[32] other scientists consider them distinct.[33]

The current asteroid belt is believed to contain only a small fraction of the mass of the primordial belt. Computer simulations suggest that the original asteroid belt may have contained mass equivalent to the Earth. Primarily because of gravitational perturbations, most of the material was ejected from the belt within about a million years of formation, leaving behind less than 0.1% of the original mass.[26] Since their formation, the size distribution of the asteroid belt has remained relatively stable: there has been no significant increase or decrease in the typical dimensions of the main-belt asteroids.[34]

The 4:1 orbital resonance with Jupiter, at a radius 2.06 AU, can be considered the inner boundary of the asteroid belt. Perturbations by Jupiter send bodies straying there into unstable orbits. Most bodies formed inside the radius of this gap were swept up by Mars (which has an aphelion at 1.67 AU) or ejected by its gravitational perturbations in the early history of the Solar System.[35] The Hungaria asteroids lie closer to the Sun than the 4:1 resonance, but are protected from disruption by their high inclination.[36]

When the asteroid belt was first formed, the temperatures at a distance of 2.7 AU from the Sun formed a "snow line" below the freezing point of water. Planetesimals formed beyond this radius were able to accumulate ice.[37] [38] In 2006 it was announced that a population of comets had been discovered within the asteroid belt beyond the snow line, which may have provided a source of water for Earth's oceans. According to some models, there was insufficient outgassing of water during the Earth's formative period to form the oceans, requiring an external source such as a cometary bombardment.[39]

# Characteristics

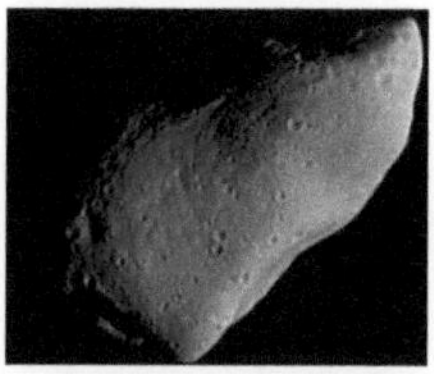

951 Gaspra, the first asteroid imaged by a spacecraft, as viewed during *Galileo*'s 1991 flyby; colors are exaggerated

Fragment of the Allende meteorite, a carbonaceous chondrite that fell to Earth in Mexico in 1969

Contrary to popular imagery, the asteroid belt is mostly empty. The asteroids are spread over such a large volume that it would be improbable to reach an asteroid without aiming carefully. Nonetheless, hundreds of thousands of asteroids are currently known, and the total number ranges in the millions or more, depending on the lower size cutoff. Over 200 asteroids are known to be larger than 100 km,[40] while a survey in the infrared wavelengths shows that the asteroid belt has 700,000 to 1.7 million asteroids with a diameter of 1 km or more.[41] The apparent magnitudes of most of the known asteroids are 11–19, with the median at about 16.[42]

The total mass of the asteroid belt is estimated to be $2.8 \times 10^{21}$ to $3.2 \times 10^{21}$ kilograms, which is just 4% of the mass of the Moon.[2] The four largest objects, Ceres, 4 Vesta, 2 Pallas, and 10 Hygiea, account for half of the belt's total mass, with almost one-third accounted for by Ceres alone.[3] [4]

## Composition

The current belt consists primarily of three categories of asteroids: C-type or carbonaceous asteroids, S-type or silicate asteroids, and M-type or metallic asteroids.

Carbonaceous asteroids, as their name suggests, are carbon-rich and dominate the belt's outer regions.[43] Together they comprise over 75% of the visible asteroids. They are more red in hue than the other asteroids and have a very low albedo. Their surface composition is similar to carbonaceous chondrite meteorites. Chemically, their spectra match the primordial composition of the early Solar System, with only the lighter elements and volatiles removed.

S-type (silicate-rich) asteroids are more common toward the inner region of the belt, within 2.5 AU of the Sun.[43] [44] The spectra of their surfaces reveal the presence of silicates and some metal, but no significant carbonaceous compounds. This indicates that their materials have been significantly modified from their primordial composition, probably through melting and reformation. They have a relatively high albedo, and form about 17% of the total asteroid population.

M-type (metal-rich) asteroids form about 10% of the total population; their spectra resemble that of iron-nickel. Some are believed to have formed from the metallic cores of differentiated progenitor bodies that were disrupted through collision. However, there are also some silicate compounds that can produce a similar appearance. For example, the large M-type asteroid 22 Kalliope does not appear to be primarily composed of metal.[45] Within the asteroid belt, the number distribution of M-type asteroids peaks at a semi-major axis of about 2.7 AU.[46] It is not yet clear whether all M-types are compositionally similar, or whether it is a label for several varieties which do not fit neatly into the main C and S classes.[47]

One mystery of the asteroid belt is the relative rarity of V-type, or basaltic asteroids.[48] Theories of asteroid formation predict that objects the size of Vesta or larger should form crusts and mantles, which would be composed mainly of basaltic rock, resulting in more than half of all asteroids being composed either of basalt or olivine. Observations, however, suggest that 99 percent of the predicted basaltic material is missing.[49] Until 2001, most basaltic bodies discovered in the asteroid belt were believed to originate from the asteroid Vesta (hence their name V-type). However, the discovery of the asteroid 1459 Magnya revealed a slightly different chemical composition from the other basaltic asteroids discovered until then, suggesting a different origin.[49] This hypothesis was reinforced by the further discovery in 2007 of two asteroids in the outer belt, 7472 Kumakiri and (10537) 1991 $RY_{16}$, with differing basaltic composition that could not have originated from Vesta. These latter two are the only V-type asteroids discovered in the outer belt to date.[48]

The temperature of the asteroid belt varies with the distance from the Sun. For dust particles within the belt, typical temperatures range from 200 K (−73 °C) at 2.2 AU down to 165 K (−108 °C) at 3.2 AU[50] However, due to rotation, the surface temperature of an asteroid can vary considerably as the sides are alternately exposed to solar radiation and then to the stellar background.

## Main-belt comets

Several otherwise unremarkable bodies in the outer belt show cometary activity. Since their orbits cannot be explained through capture of classical comets, it is thought that many of the outer asteroids may be icy, with the ice occasionally exposed to sublimation through small impacts. Main-belt comets may have been a major source of the Earth's oceans, since the deuterium-hydrogen ratio is too low for classical comets to have been the principal source.[51]

## Orbits

Most asteroids within the asteroid belt have orbital eccentricities of less than 0.4, and an inclination of less than 30°. The orbital distribution of the asteroids reaches a maximum at an eccentricity of around 0.07 and an inclination below 4°.[42] Thus while a typical asteroid has a relatively circular orbit and lies near the plane of the ecliptic, some asteroid orbits can be highly eccentric or travel well outside the ecliptic plane.

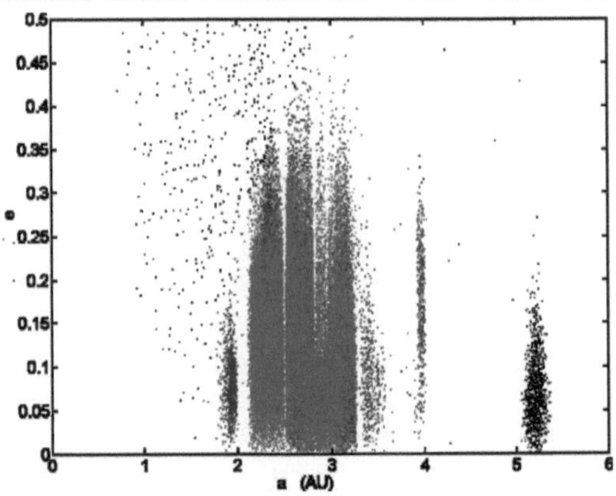

The asteroid belt (showing eccentricities), with the asteroid belt in red and blue ("core" region in red)

Sometimes, the term *main belt* is used to refer only to the more compact "core" region where the greatest concentration of bodies is found. This lies between the strong 4:1 and 2:1 Kirkwood gaps at 2.06 and 3.27 AU, and at orbital eccentricities less than roughly 0.33, along with orbital inclinations below about 20°. This "core" region contains approximately 93.4% of all numbered minor planets within the Solar System.[52]

### Kirkwood gaps

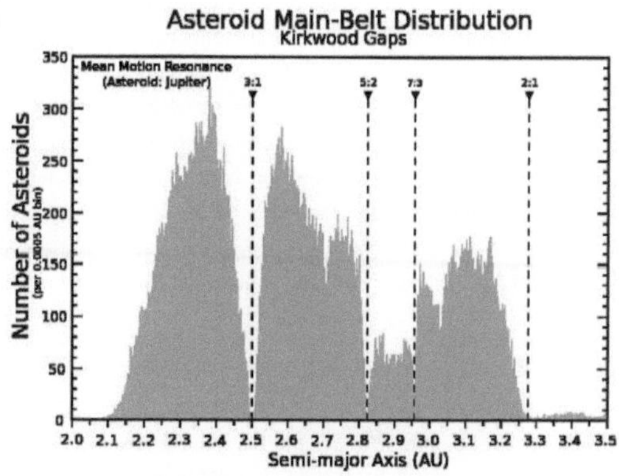


This chart shows the distribution of asteroid semi-major axes in the "core" of the asteroid belt. Black arrows point to the Kirkwood gaps, where orbital resonances with Jupiter destabilize orbits.

The semi-major axis of an asteroid is used to describe the dimensions of its orbit around the Sun, and its value determines the minor planet's orbital period. In 1866, Daniel Kirkwood announced the discovery of gaps in the distances of these bodies' orbits from the Sun. They were located at positions where their period of revolution about the Sun was an integer fraction of Jupiter's orbital period. Kirkwood proposed that the gravitational perturbations of the planet led to the removal of asteroids from these orbits.[53]

When the mean orbital period of an asteroid is an integer fraction of the orbital period of Jupiter, a mean-motion resonance with the gas giant is created that is sufficient to perturb an asteroid to new orbital elements. Asteroids that become located in the gap orbits (either primordially because of the migration of Jupiter's orbit,[54] or due to prior perturbations or collisions) are gradually nudged into different, random orbits with a larger or smaller semi-major axis.

The gaps are not seen in a simple snapshot of the locations of the asteroids at any one time because asteroid orbits are elliptical, and many asteroids still cross through the radii corresponding to the gaps. The actual spatial density of asteroids in these gaps does not differ significantly from the neighboring regions.[55]

The main gaps occur at the 3:1, 5:2, 7:3, and 2:1 mean-motion resonances with Jupiter. An asteroid in the 3:1 Kirkwood gap would orbit the Sun three times for each Jovian orbit, for instance. Weaker resonances occur at other semi-major axis values, with fewer asteroids found than nearby. (For example, an 8:3 resonance for asteroids with a semi-major axis of 2.71 AU.)[56]

The main or core population of the asteroid belt is sometimes divided into three zones, based on the most prominent Kirkwood gaps. Zone I lies between the 4:1 resonance (2.06 AU) and 3:1 resonance (2.5 AU) Kirkwood gaps. Zone II continues from the end of Zone I out to the 5:2 resonance gap (2.82 AU). Zone III extends from the outer edge of Zone II to the 2:1 resonance gap (3.28 AU).[57]

The asteroid belt may also be divided into the inner and outer belts, with the inner belt formed by asteroids orbiting nearer to Mars than the 3:1 Kirkwood gap (2.5 AU), and the outer belt formed by those asteroids closer to Jupiter's orbit. (Some authors subdivide the inner and outer belts at the 2:1 resonance gap (3.3 AU), while others suggest inner, middle, and outer belts.)

## Collisions

The zodiacal light, created in part by dust from collisions in the asteroid belt

The high population of the asteroid belt makes for a very active environment, where collisions between asteroids occur frequently (on astronomical time scales). Collisions between main-belt bodies with a mean radius of 10 km are expected to occur about once every 10 million years.[58] A collision may fragment an asteroid into numerous smaller pieces (leading to the formation of a new asteroid family). Conversely, collisions that occur at low relative speeds may also join two asteroids. After more than 4 billion years of such processes, the members of the asteroid belt now bear little resemblance to the original population.

Along with the asteroid bodies, the asteroid belt also contains bands of dust with particle radii of up to a few hundred micrometres. This fine material is produced, at least in part, from collisions between asteroids, and by the impact of micrometeorites upon the asteroids. Due to the Poynting-Robertson effect, the pressure of solar radiation causes this dust to slowly spiral inward toward the Sun.[59]

The combination of this fine asteroid dust, as well as ejected cometary material, produces the zodiacal light. This faint auroral glow can be viewed at night extending from the direction of the Sun along the plane of the ecliptic. Particles that produce the visible zodiacal light average about 40 μm in radius. The typical lifetimes of such particles are about 700,000 years. Thus, to maintain the bands of dust, new particles must be steadily produced within the asteroid belt.[59]

### Meteorites

Some of the debris from collisions can form meteoroids that enter the Earth's atmosphere.[60] Of the 50,000 meteorites found on Earth to date, 99.8 percent are believed to have originated in the asteroid belt.[61] A September 2007 study has suggested that a large-body collision undergone by the asteroid 298 Baptistina sent a number of fragments into the inner Solar System. The impacts of these fragments are believed to have created both Tycho crater on the Moon and Chicxulub crater in Mexico, the relict of the massive impact which is believed to have triggered the extinction of the dinosaurs 65 million years ago.[62]

## Families and groups

In 1918, the Japanese astronomer Kiyotsugu Hirayama noticed that the orbits of some of the asteroids had similar parameters, forming families or groups.[63]

Approximately one-third of the asteroids in the asteroid belt are members of an asteroid family. These share similar orbital elements, such as semi-major axis, eccentricity, and orbital inclination as well as similar spectral features, all of which indicate a common origin in the breakup of a larger body. Graphical displays of these elements, for members of the asteroid belt, show concentrations indicating the presence of an asteroid family. There are about 20–30 associations that are almost certainly asteroid families. Additional groupings have been found that are less certain. Asteroid families can be confirmed when the members display common spectral features.[64] Smaller associations of asteroids are called groups or clusters.

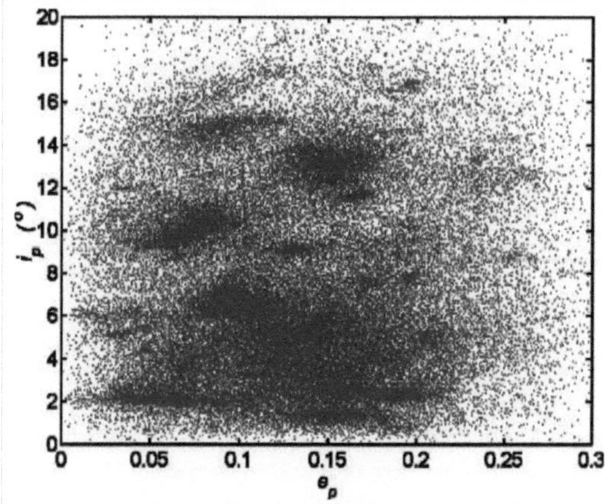


This plot of orbital inclination ($i_p$) versus eccentricity ($e_p$) for the numbered main-belt asteroids clearly shows clumpings representing asteroid families.

Some of the most prominent families in the asteroid belt (in order of increasing semi-major axes) are the Flora, Eunoma, Koronis, Eos, and Themis families.[46] The Flora family, one of the largest with more than 800 known members, may have formed from a collision less than a billion years ago.[65] The largest asteroid to be a true member of a family (as opposed to an interloper in the case of Ceres with the Gefion family) is 4 Vesta. The Vesta family is believed to have formed as the result of a crater-forming impact on Vesta. Likewise, the HED meteorites may also have originated from Vesta as a result of this collision.[66]

Three prominent bands of dust have been found within the asteroid belt. These have similar orbital inclinations as the Eos, Koronis, and Themis asteroid families, and so are possibly associated with those groupings.[67]

### Periphery

Skirting the inner edge of the belt (ranging between 1.78 and 2.0 AU, with a mean semi-major axis of 1.9 AU) is the Hungaria family of minor planets. They are named after the main member, 434 Hungaria; the group contains at least 52 named asteroids. The Hungaria group is separated from the main body by the 4:1 Kirkwood gap and their orbits have a high inclination. Some members belong to the Mars-crossing category of asteroids, and gravitational perturbations by Mars are likely a factor in reducing the total population of this group.[68]

Another high-inclination group in the inner part of the asteroid belt is the Phocaea family. These are composed primarily of S-type asteroids, whereas the neighboring Hungaria family includes some E-types.[69] The Phocaea family orbit between 2.25 and 2.5 AU from the Sun.

Skirting the outer edge of the asteroid belt is the Cybele group, orbiting between 3.3 and 3.5 AU. These have a 7:4 orbital resonance with Jupiter. The Hilda family orbit between 3.5 and 4.2 AU, and have relatively circular orbits and a stable 3:2 orbital resonance with Jupiter. There are few asteroids beyond 4.2 AU, until Jupiter's orbit. Here the two families of Trojan asteroids can be found, which, at least for objects larger than 1 km, are approximately as numerous as the asteroids of the asteroid belt.[70]

### New families

Some asteroid families have formed recently, in astronomical terms. The Karin Cluster apparently formed about 5.7 million years ago from a collision with a 33 km radius progenitor asteroid.[71] The Veritas family formed about 8.3 million years ago; evidence includes interplanetary dust recovered from ocean sediment.[72]

More recently, the Datura cluster appears to have formed about 450 thousand years ago from a collision with a main-belt asteroid. The age estimate is based on the probability of the members having their current orbits, rather than from any physical evidence. However, this cluster may have been a source for some zodiacal dust material.[73] Other recent cluster formations, such as the Iannini cluster (*circa* 1–5 million years ago), may have provided additional sources of this asteroid dust.[74]

## Exploration

The first spacecraft to traverse the asteroid belt was Pioneer 10, which entered the region on July 16, 1972. At the time there was some concern that the debris in the belt would pose a hazard to the spacecraft, but it has since been safely traversed by 11 Earth-based craft without incident. Pioneer 11, Voyagers 1 and 2 and Ulysses passed through the belt without imaging any asteroids. Galileo imaged the asteroid 951 Gaspra in 1991 and 243 Ida in 1993, NEAR imaged 253 Mathilde in 1997, Cassini imaged 2685 Masursky in 2000, Stardust imaged 5535 Annefrank in 2002, New Horizons imaged 132524 APL in 2006, Rosetta imaged 2867 Šteins in 2008, and Dawn has been orbiting Vesta since July 2011.[75] Due to the low density of materials within the belt, the odds of a probe running into an asteroid are now estimated at less than one in a billion.[76]

Artist's concept of the Dawn spacecraft with Vesta (left) and Ceres (right)

Most belt asteroids imaged to date have come from brief flyby opportunities by probes headed for other targets. Only the Dawn, NEAR and Hayabusa missions have studied asteroids for a protracted period in orbit and at the surface. Dawn explored Vesta from July 2011 to September 2012, and is currently en route to Ceres for a 2015 rendezvous. If the probe is still operational after examining Ceres, an extended mission could allow additional exploration, possibly a flyby of Pallas.[77]

## See also

- Asteroid mining
- Asteroids in astrology
- Asteroids in fiction
- Colonization of the asteroids
- Debris disk
- Kuiper belt

## References


[1] Krasinsky, G. A.; Pitjeva, E. V.; Vasilyev, M. V.; Yagudina, and E. I. (July 2002). "Hidden Mass in the Asteroid Belt". *Icarus* **158** (1): 98–105. Bibcode 2002Icar..158...98K. doi:10.1006/icar.2002.6837.

[2] Pitjeva, E. V. (2005). "High-Precision Ephemerides of Planets—EPM and Determination of Some Astronomical Constants" (http://iau-comm4.jpl.nasa.gov/EPM2004.pdf) (PDF). *Solar System Research* **39** (3): 176. Bibcode 2005SoSyR..39..176P. doi:10.1007/s11208-005-0033-2. .

[3] For recent estimates of the masses of Ceres, 4 Vesta, 2 Pallas and 10 Hygiea, see the references in the infoboxes of their respective articles.

[4] Yeomans, Donald K. (July 13, 2006). "JPL Small-Body Database Browser" (http://ssd.jpl.nasa.gov/sbdb.cgi). NASA JPL. Archived (http://web.archive.org/web/20100929043420/http://ssd.jpl.nasa.gov/sbdb.cgi) from the original on 29 September 2010. . Retrieved 2010-09-27.

[5] Hilton, J. (2001). "When Did the Asteroids Become Minor Planets?" (http://www.usno.navy.mil/USNO/astronomical-applications/astronomical-information-center/minor-planets). *US Naval Observatory (USNO)*. . Retrieved 2007-10-01.

[6] "Dawn: A Journey to the Beginning of the Solar System" (http://www-ssc.igpp.ucla.edu/dawn/background.html). *Space Physics Center: UCLA*. 2005. . Retrieved 2007-11-03.

[7] Hoskin, Michael. "Bode's Law and the Discovery of Ceres" (http://www.astropa.unipa.it/HISTORY/hoskin.html). *Churchill College, Cambridge*. . Retrieved 2010-07-12.

[8] Linda T. Elkins-Tanton, *Asteroids, Meteorites, and Comets*, 2010:10

[9] "Call the police! The story behind the discovery of the asteroids". *Astronomy Now* (June 2007): 60–61.

[10] Pogge, Richard (2006). "An Introduction to Solar System Astronomy: Lecture 45: Is Pluto a Planet?" (http://www.astronomy.ohio-state.edu/~pogge/Ast161/Unit6/dwarfs.html). *An Introduction to Solar System Astronomy*. Ohio State University. . Retrieved 2007-11-11.

[11] Harper, Douglas (2010). "Asteroid" (http://www.etymonline.com/index.php?search=asteroid&searchmode=none). *Online Etymology Dictionary*. Etymology Online. . Retrieved 2011-04-15.

[12] DeForest, Jessica (2000). "Greek and Latin Roots" (http://www.msu.edu/~defores1/gre/roots/gre_rts_afx2.htm). Michigan State University. Archived (http://web.archive.org/web/20070812155525/http://www.msu.edu/~defores1/gre/roots/gre_rts_afx2.htm) from the original on 12 August 2007. . Retrieved 2007-07-25.

[13] Cunningham, Clifford (1984). "William Hershel and the First Two Asteroids". *Dance Hall Observatory, Ontario* **11**: 3. Bibcode 1984MPBu...11....3C.

[14] Staff (2002). "Astronomical Serendipity" (http://dawn.jpl.nasa.gov/DawnCommunity/flashbacks/fb_06.asp). NASA JPL. . Retrieved 2007-04-20.

[15] "Is it a coincidence that most of the planets fall within the Titius-Bode law's boundaries?" (http://www.astronomy.com/asy/default.aspx?c=a&id=4494). *astronomy.com*. . Retrieved 2007-10-16.

[16] von Humboldt, Alexander (1850). *Cosmos: A Sketch of a Physical Description of the Universe*. **1**. Harper & Brothers, New York (NY). p. 44. ISBN 0-8018-5503-9.

[17] Mann, Robert James (1852). *A Guide to the Knowledge of the Heavens*. Jarrold. p. 171. and 1853, p. 216

[18] "Further Investigation relative to the form, the magnitude, the mass, and the orbit of the Asteroid Planets" (http://books.google.com/?id=hhQAAAAAMAAJ&pg=PA191&dq=asteroid+belt). *The Edinburgh New Philosophical Journal* **5**: 191. January–April 1857. .: "[Professor Peirce] then observed that the analogy between the ring of Saturn and the belt of the asteroids was worthy of notice."

[19] Hughes, David W. (2007). "A Brief History of Asteroid Spotting" (http://www.open2.net/sciencetechnologynature/planetsbeyond/asteroids/history.html). BBC. . Retrieved 2007-04-20.

[20] Moore, Patrick; Rees, Robin (2011). *Patrick Moore's Data Book of Astronomy* (2nd ed.). Cambridge University Press. p. 156. ISBN 0-521-89935-4.

[21] Manley, Scott (August 25, 2010). *Asteroid Discovery from 1980 to 2010* (http://www.youtube.com/watch?v=S_d-gs0WoUw). . Retrieved 2011-04-15.

[22] "MPC Archive Statistics" (http://www.minorplanetcenter.org/iau/lists/ArchiveStatistics.html). IAU Minor Planet Center. . Retrieved 2011-04-04.

[23] "A Brief History of Asteroid Spotting" (http://www.open2.net/sciencetechnologynature/planetsbeyond/asteroids/history.html). *Open2.net*. . Retrieved 2007-05-15.

[24] Masetti, M.; and Mukai, K. (December 1, 2005). "Origin of the Asteroid Belt" (http://imagine.gsfc.nasa.gov/docs/ask_astro/answers/980810a.html). NASA Goddard Spaceflight Center. . Retrieved 2007-04-25.

[25] Watanabe, Susan (July 20, 2001). "Mysteries of the Solar Nebula" (http://www.jpl.nasa.gov/news/features.cfm?feature=520). NASA. . Retrieved 2007-04-02.

[26] Petit, J.-M.; Morbidelli, A.; and Chambers, J. (2001). "The Primordial Excitation and Clearing of the Asteroid Belt" (http://www.gps.caltech.edu/classes/ge133/reading/asteroids.pdf) (PDF). *Icarus* **153** (2): 338–347. Bibcode 2001Icar..153..338P. doi:10.1006/icar.2001.6702. Archived (http://web.archive.org/web/20070221085835/http://www.gps.caltech.edu/classes/ge133/reading/asteroids.pdf) from the original on 21 February 2007. . Retrieved 2007-03-22.

[27] Edgar, R.; and Artymowicz, P. (2004). "Pumping of a Planetesimal Disc by a Rapidly Migrating Planet" (http://www.astro.su.se/~pawel/edgar+artymowicz.pdf) (PDF). *Monthly Notices of the Royal Astronomical Society* **354** (3): 769–772. arXiv:astro-ph/0409017. Bibcode 2004MNRAS.354..769E. doi:10.1111/j.1365-2966.2004.08238.x. . Retrieved 2007-04-16.

[28] Scott, E. R. D. (March 13–17, 2006). "Constraints on Jupiter's Age and Formation Mechanism and the Nebula Lifetime from Chondrites and Asteroids" (http://adsabs.harvard.edu/abs/2006LPI....37.2367S). *Proceedings 37th Annual Lunar and Planetary Science Conference*. League City, Texas: Lunar and Planetary Society. . Retrieved 2007-04-16.

[29] Taylor, G. J.; Keil, K.; McCoy, T.; Haack, H.; and Scott, E. R. D.; Keil; McCoy; Haack; Scott (1993). "Asteroid differentiation – Pyroclastic volcanism to magma oceans". *Meteoritics* **28** (1): 34–52. Bibcode 1993Metic..28...34T.

[30] Kelly, Karen (2007). "U of T researchers discover clues to early solar system" (http://webapps.utsc.utoronto.ca/ose/story.php?id=665). *University of Toronto*. . Retrieved 2010-07-12.

[31] Clark, B. E.; Hapke, B.; Pieters, C.; Britt, D.; Hapke; Pieters; Britt (2002). "Asteroid Space Weathering and Regolith Evolution". *University of Arizona*: 585. Bibcode 2002aste.conf..585C. Gaffey, Michael J. (1996). "The Spectral and Physical Properties of Metal in Meteorite Assemblages: Implications for Asteroid Surface Materials". *Icarus (ISSN 0019-1035)* **66** (3): 468. Bibcode 1986Icar...66..468G. doi:10.1016/0019-1035(86)90086-2. Keil, K. (2000). "Thermal alteration of asteroids: evidence from meteorites" (http://www.ingentaconnect.com/content/els/00320633/2000/00000048/00000010/art00054). *Planetary and Space Science*. . Retrieved 2007-11-08. Baragiola, R. A.; Duke, C. A.; Loeffler, M.; McFadden, L. A.; and Sheffield, J.; Duke; Loeffler; McFadden; Sheffield (2003). "Impact of ions and micrometeorites on mineral surfaces: Reflectance changes and production of atmospheric species in airless solar system bodies". *EGS - AGU - EUG Joint Assembly*: 7709. Bibcode 2003EAEJA.....7709B.

[32] Chapman, C. R.; Williams, J. G.; Hartmann, W. K. (1978). "The asteroids". *Annual review of astronomy and astrophysics* **16**: 33–75. Bibcode 1978ARA&A..16...33C. doi:10.1146/annurev.aa.16.090178.000341.

[33] Kracher, A. (2005). "Asteroid 433 Eros and partially differentiated planetesimals: bulk depletion versus surface depletion of sulfur" (http://www.cosis.net/abstracts/EGU05/03788/EGU05-J-03788.pdf) (PDF). *Ames Laboratory*. Archived (http://web.archive.org/web/20071128200221/http://www.cosis.net/abstracts/EGU05/03788/EGU05-J-03788.pdf) from the original on 28 November 2007. . Retrieved 2007-11-08.

[34] Stiles, Lori (September 15, 2005). "Asteroids Caused the Early Inner Solar System Cataclysm" (http://uanews.org/cgi-bin/WebObjects/UANews.woa/7/wa/SRStoryDetails?ArticleID=11692). University of Arizona News. . Retrieved 2007-04-18.

[35] Alfvén, H.; Arrhenius, G. (1976). "The Small Bodies" (http://history.nasa.gov/SP-345/ch4.htm). *SP-345 Evolution of the Solar System*. NASA. Archived (http://web.archive.org/web/20070513081833/http://history.nasa.gov/SP-345/ch4.htm) from the original on 13 May 2007. . Retrieved 2007-04-12.

[36] Spratt, Christopher E. (April 1990). "The Hungaria group of minor planets". *Journal of the Royal Astronomical Society of Canada* **84**: 123–131. Bibcode 1990JRASC..84..123S.

[37] Lecar, M.; Podolak, M.; Sasselov, D.; Chiang, E. (2006). "Infrared cirrus – New components of the extended infrared emission". *The Astrophysical Journal* **640** (2): 1115–1118. arXiv:astro-ph/0602217. Bibcode 2006ApJ...640.1115L. doi:10.1086/500287.

[38] Berardelli, Phil (March 23, 2006). "Main-Belt Comets May Have Been Source Of Earths Water" (http://www.spacedaily.com/reports/Main_Belt_Comets_May_Have_Been_Source_Of_Earths_Water.html). Space Daily. Archived (http://web.archive.org/web/20071011171815/http://spacedaily.com/reports/Main_Belt_Comets_May_Have_Been_Source_Of_Earths_Water.html) from the original on 11 October 2007. . Retrieved 2007-10-27.

[39] Lakdawalla, Emily (April 28, 2006). "Discovery of a Whole New Type of Comet" (http://www.planetary.org/blog/article/00000551/). The Planetary Society. Archived (http://web.archive.org/web/20070501211319/http://www.planetary.org/blog/article/00000551/) from the original on 1 May 2007. . Retrieved 2007-04-20.

[40] Yeomans, Donald K. (April 26, 2007). "JPL Small-Body Database Search Engine" (http://ssd.jpl.nasa.gov/sbdb_query.cgi). NASA JPL. . Retrieved 2007-04-26. – search for asteroids in the main belt regions with a diameter >100.

[41] Tedesco, E. F.; and Desert, F.-X. (2002). "The Infrared Space Observatory Deep Asteroid Search". *The Astronomical Journal* **123** (4): 2070–2082. Bibcode 2002AJ....123.2070T. doi:10.1086/339482.

[42] Williams, Gareth (September 25, 2010). "Distribution of the Minor Planets" (http://www.minorplanetcenter.org/iau/lists/MPDistribution.html). Minor Planets Center. . Retrieved 2010-10-27.

[43] Wiegert, P.; Balam, D.; Moss, A.; Veillet, C.; Connors, M.; and Shelton, I. (2007). "Evidence for a Color Dependence in the Size Distribution of Main-Belt Asteroids" (http://astro.uwo.ca/~wiegert/papers/2007AJ.133.1609.pdf). *The Astronomical Journal* **133** (4): 1609–1614. arXiv:astro-ph/0611310. Bibcode 2007AJ....133.1609W. doi:10.1086/512128. . Retrieved 2008-09-06.

[44] Clark, B. E. (1996). "New News and the Competing Views of Asteroid Belt Geology". *Lunar and Planetary Science* **27**: 225–226. Bibcode 1996LPI....27..225C.

[45] Margot, J. L.; and Brown, M. E. (2003). "A Low-Density M-type Asteroid in the Main Belt". *Science* **300** (5627): 1939–1942. Bibcode 2003Sci...300.1939M. doi:10.1126/science.1085844. PMID 12817147.

[46] Lang, Kenneth R. (2003). "Asteroids and meteorites" (http://ase.tufts.edu/cosmos/print_images.asp?id=15). NASA's Cosmos. . Retrieved 2007-04-02.

[47] Mueller, M.; Harris, A. W.; Delbo, M.; and the MIRSI Team; Harris; Delbo (2005). "21 Lutetia and other M-types: Their sizes, albedos, and thermal properties from new IRTF measurements". *Bulletin of the American Astronomical Society* **37**: 627. Bibcode 2005DPS....37.0702M.

[48] Duffard, R. D.; Roig, F. (July 14–18, 2008). "Two New Basaltic Asteroids in the Main Belt?". *Asteroids, Comets, Meteors 2008*. Baltimore, Maryland. arXiv:0704.0230. Bibcode 2008LPICo1405.8154D.

[49] Than, Ker (2007). "Strange Asteroids Baffle Scientists" (http://www.space.com/scienceastronomy/070821_basalt_asteroid.html). *space.com*. . Retrieved 2007-10-14.

[50] Low, F. J.; *et al.* (1984). "Infrared cirrus – New components of the extended infrared emission". *Astrophysical Journal, Part 2 – Letters to the Editor* **278**: L19–L22. Bibcode 1984ApJ...278L..19L. doi:10.1086/184213.

[51] "Interview with David Jewitt" (http://www.youtube.com/watch?v=B1W4NTmI5Bk). Youtube.com. 2007-01-05. . Retrieved 2011-05-21.

[52] This value was obtained by a simple count up of all bodies in that region using data for 120437 numbered minor planets from the Minor Planet Center orbit database (http://www.minorplanetcenter.org/iau/MPCORB.html), dated February 8, 2006.

[53] Fernie, J. Donald (1999). "The American Kepler" (http://www.americanscientist.org/issues/pub/1999/9/the-american-kepler/2). *The Americal Scientist* **87** (5): 398. . Retrieved 2007-02-04.

[54] Liou, Jer-Chyi; and Malhotra, Renu (1997). "Depletion of the Outer Asteroid Belt" (http://www.sciencemag.org/cgi/content/full/275/5298/375). *Science* **275** (5298): 375–377. Bibcode 1997Sci...275..375L. doi:10.1126/science.275.5298.375. PMID 8994031. . Retrieved 2007-08-01.

[55] McBride, N.; and Hughes, D. W.; Hughes (1990). "The spatial density of asteroids and its variation with asteroidal mass". *Monthly Notices of the Royal Astronomical Society* **244**: 513–520. Bibcode 1990MNRAS.244..513M.

[56] Ferraz-Mello, S. (June 14–18, 1993). "Kirkwood Gaps and Resonant Groups" (http://adsabs.harvard.edu/abs/1994IAUS..160..175F). *proceedings of the 160th International Astronomical Union*. Belgirate, Italy: Kluwer Academic Publishers. pp. 175–188. . Retrieved 2007-03-28.

[57] Klacka, Jozef (1992). "Mass distribution in the asteroid belt". *Earth, Moon, and Planets* **56** (1): 47–52. Bibcode 1992EM&P...56...47K. doi:10.1007/BF00054599.

[58] Backman, D. E. (March 6, 1998). "Fluctuations in the General Zodiacal Cloud Density" (http://astrobiology.arc.nasa.gov/workshops/zodiac/backman/backman_toc.html). *Backman Report*. NASA Ames Research Center. . Retrieved 2007-04-04.

[59] Reach, William T. (1992). "Zodiacal emission. III – Dust near the asteroid belt". *Astrophysical Journal* **392** (1): 289–299. Bibcode 1992ApJ...392..289R. doi:10.1086/171428.

[60] Kingsley, Danny (May 1, 2003). "Mysterious meteorite dust mismatch solved" (http://abc.net.au/science/news/stories/s843594.htm). ABC Science. . Retrieved 2007-04-04.

[61] "Meteors and Meteorites" (http://www.nasa.gov/pdf/145945main_Meteors.Meteorites.Lithograph.pdf). NASA. . Retrieved 2012-01-12.

[62] "Breakup event in the main asteroid belt likely caused dinosaur extinction 65 million years ago" (http://www.physorg.com/news108218928.html). *Southwest Research Institute*. 2007. Archived (http://web.archive.org/web/20071011151507/http://physorg.com/news108218928.html) from the original on 11 October 2007. . Retrieved 2007-10-14.

[63] Hughes, David W. (2007). "Finding Asteroids In Space" (http://www.open2.net/sciencetechnologynature/planetsbeyond/asteroids/finding.html). BBC. . Retrieved 2007-04-20.

[64] Lemaitre, Anne (August 31-September 4, 2004). "Asteroid family classification from very large catalogues" (http://adsabs.harvard.edu/abs/2005dpps.conf..135L). *Proceedings Dynamics of Populations of Planetary Systems*. Belgrade, Serbia and Montenegro: Cambridge University Press. pp. 135–144. . Retrieved 2007-04-15.

[65] Martel, Linda M. V. (March 9, 2004). "Tiny Traces of a Big Asteroid Breakup" (http://www.psrd.hawaii.edu/Mar04/fossilMeteorites.html). Planetary Science Research Discoveries. Archived (http://web.archive.org/web/20070401115157/http://www.psrd.hawaii.edu/Mar04/fossilMeteorites.html) from the original on 1 April 2007. . Retrieved 2007-04-02.

[66] Drake, Michael J. (2001). "The eucrite/Vesta story". *Meteoritics & Planetary Science* **36** (4): 501–513. Bibcode 2001M&PS...36..501D. doi:10.1111/j.1945-5100.2001.tb01892.x.

[67] Love, S. G.; and Brownlee, D. E. (1992). "The IRAS dust band contribution to the interplanetary dust complex – Evidence seen at 60 and 100 microns". *Astronomical Journal* **104** (6): 2236–2242. Bibcode 1992AJ....104.2236L. doi:10.1086/116399.

[68] Spratt, Christopher E. (1990). "The Hungaria group of minor planets". *Journal of the Royal Astronomical Society of Canada* **84** (2): 123–131. Bibcode 1990JRASC..84..123S.

[69] Carvano, J. M.; Lazzaro, D.; Mothé-Diniz, T.; Angeli, C. A.; and Florczak, M. (2001). "Spectroscopic Survey of the Hungaria and Phocaea Dynamical Groups". *Icarus* **149** (1): 173–189. Bibcode 2001Icar..149..173C. doi:10.1006/icar.2000.6512.

[70] Dymock, Roger (2010). *Asteroids and Dwarf Planets and How to Observe Them* (http://books.google.com/books?id=vQcAnwt_87sC&pg=PA24). Astronomers' Observing Guides. Springer. p. 24. ISBN 1-4419-6438-X. . Retrieved 2011-04-04.

[71] Nesvorný, David; *et al.* (August 2006). "Karin cluster formation by asteroid impact". *Icarus* **183** (2): 296–311. Bibcode 2006Icar..183..296N. doi:10.1016/j.icarus.2006.03.008.

[72] McKee, Maggie (January 18, 2006). "Eon of dust storms traced to asteroid smash" (http://space.newscientist.com/channel/solar-system/comets-asteroids/dn8603). New Scientist Space. . Retrieved 2007-04-15.

[73] Nesvorný, D.; Vokrouhlick, D.; and Bottke, W. F. (2006). "The Breakup of a Main-Belt Asteroid 450 Thousand Years Ago" (http://www.sciencemag.org/cgi/content/full/312/5779/1490). *Science* **312** (5779): 1490. Bibcode 2006Sci...312.1490N. doi:10.1126/science.1126175. PMID 16763141. . Retrieved 2007-04-15.

[74] Nesvorný, D.; Bottke, W. F.; Levison, H. F.; and Dones, L. (2003). "Recent Origin of the Solar System Dust Bands" (http://iopscience.iop.org/0004-637X/591/1/486/pdf/0004-637X_591_1_486.pdf). *The Astrophysical Journal* **591** (1): 486–497. Bibcode 2003ApJ...591..486N. doi:10.1086/374807. . Retrieved 2007-04-15.

[75] Barucci, M. A.; Fulchignoni, M.; and Rossi, A. (2007). "Rosetta Asteroid Targets: 2867 Steins and 21 Lutetia". *Space Science Reviews* **128** (1–4): 67–78. Bibcode 2007SSRv..128...67B. doi:10.1007/s11214-006-9029-6.

[76] Stern, Alan (June 2, 2006). "New Horizons Crosses The Asteroid Belt" (http://www.spacedaily.com/reports/New_Horizons_Crosses_The_Asteroid_Belt.html). Space Daily. . Retrieved 2007-04-14.
[77] Staff (April 10, 2007). "Dawn Mission Home Page" (http://dawn.jpl.nasa.gov/). NASA JPL. Archived (http://web.archive.org/web/20070411053350/http://dawn.jpl.nasa.gov/) from the original on 11 April 2007. . Retrieved 2007-04-14.

## Further reading

- Elkins-Tanton, Linda T. (2006). *Asteroids, Meteorites, and Comets* (First ed.). New York: Chelsea House. ISBN 0-8160-5195-X.

## External links

- Asteroid Discovery from 1980 to 2010 (http://www.youtube.com/watch?v=S_d-gs0WoUw)
- Arnett, William A. (February 26, 2006). "Asteroids" (http://www.nineplanets.org/asteroids.html). The Nine Planets. Archived (http://web.archive.org/web/20070418220152/http://www.nineplanets.org/asteroids.html) from the original on 18 April 2007. Retrieved 2007-04-20.
- Asteroids Page (http://solarsystem.nasa.gov/planets/profile.cfm?Object=Asteroids) at NASA's Solar System Exploration (http://solarsystem.nasa.gov)
- Cain, Fraser. "The Asteroid Belt" (http://www.astronomycast.com/astronomy/episode-55-the-asteroid-belt/). Universe Today. Archived (http://web.archive.org/web/20080307033324/http://www.astronomycast.com/astronomy/episode-55-the-asteroid-belt/) from the original on 7 March 2008. Retrieved 2008-04-01.
- Hsieh, Henry H. (March 1, 2006). "Main-Belt Comets" (http://star.pst.qub.ac.uk/~hhh/mbcs.shtml). University of Hawaii. Retrieved 2007-04-20.
- "Main Asteroid Belt" (http://www.solstation.com/stars/asteroid.htm). Sol Company. Archived (http://web.archive.org/web/20070515163455/http://www.solstation.com/stars/asteroid.htm) from the original on 15 May 2007. Retrieved 2007-04-20.
- Munsell, Kirk (September 16, 2005). "Asteroids: Overview" (http://solarsystem.nasa.gov/planets/profile.cfm?Object=Asteroids). NASA's Solar System Exploration (http://solarsystem.nasa.gov). Archived (http://web.archive.org/web/20070524185042/http://solarsystem.nasa.gov/planets/profile.cfm?Object=Asteroids) from the original on 24 May 2007. Retrieved 2007-05-26.
- Plots of eccentricity vs. semi-major axis (http://burro.astr.cwru.edu/stu/media/asteroid_all_axisvecc.jpg) and inclination vs. semi-major axis (http://burro.astr.cwru.edu/stu/media/asteroid_all_axisvincl.jpg) at Asteroid Dynamic Site
- Staff (October 31, 2006). "Asteroids" (http://nssdc.gsfc.nasa.gov/planetary/planets/asteroidpage.html). NASA. Archived (http://web.archive.org/web/20070411201013/http://nssdc.gsfc.nasa.gov/planetary/planets/asteroidpage.html) from the original on 11 April 2007. Retrieved 2007-04-20.
- Staff (2007). "Space Topics: Asteroids and Comets" (http://www.planetary.org/explore/topics/asteroids_and_comets/facts.html). The Planetary Society. Archived (http://web.archive.org/web/20070428143402/http://www.planetary.org/explore/topics/asteroids_and_comets/facts.html) from the original on 28 April 2007. Retrieved 2007-04-20.

# Asteroid

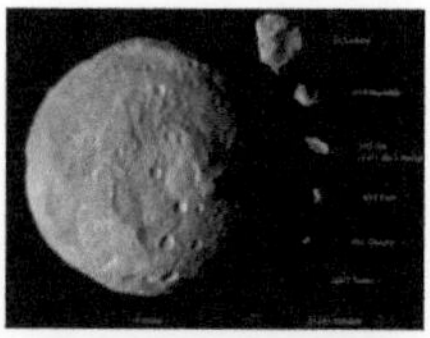

A composite image, to scale, of the asteroids which have been imaged at high resolution. As of 2011 they are, from largest to smallest: 4 Vesta, 21 Lutetia, 253 Mathilde, 243 Ida and its moon Dactyl, 433 Eros, 951 Gaspra, 2867 Šteins, 25143 Itokawa.

The largest asteroid in the previous image, Vesta (left), with Ceres (center) and Earth's Moon (right) shown to scale.

**Asteroids** are a class of small Solar System bodies in orbit around the Sun. They have also been called **planetoids**, especially the larger ones. These terms have historically been applied to any astronomical object orbiting the Sun that did not show the disk of a planet and was not observed to have the characteristics of an active comet, but as small objects in the outer Solar System were discovered, their volatile-based surfaces were found to more closely resemble comets, and so were often distinguished from traditional asteroids.[1] Thus the term *asteroid* has come increasingly to refer specifically to the small bodies of the inner Solar System out to the orbit of Jupiter, which are usually rocky or metallic. They are grouped with the outer bodies—centaurs, Neptune trojans, and trans-Neptunian objects—as minor planets, which is the term preferred in astronomical circles.[2] This article will restrict the use of the term "asteroid" to the minor planets of the inner Solar System.

There are millions of asteroids, many thought to be the shattered remnants of planetesimals, bodies within the young Sun's solar nebula that never grew large enough to become planets.[3] A large majority of known asteroids orbit in the asteroid belt between the orbits of Mars and Jupiter or co-orbital with Jupiter (the Jupiter Trojans). However, other orbital families exist with significant populations, including the near-Earth asteroids. Individual asteroids are classified by their characteristic spectra, with the majority falling into three main groups: C-type, S-type, and M-type. These were named after and are generally identified with carbon-rich, stony, and metallic compositions, respectively.

## Naming

A newly discovered asteroid is given a provisional designation (such as 2002 $AT_4$) consisting of the year of discovery and an alphanumeric code indicating the half-month of discovery and the sequence within that half-month. Once an asteroid's orbit has been confirmed, it is given a number, and later may also be given a name (e.g. 433 Eros). The formal naming convention uses parentheses around the number (e.g. (433) Eros), but dropping the parentheses is quite common. Informally, it is common to drop the number altogether, or to drop it after the first mention when a name is repeated in running text.

## Symbols

The first asteroids to be discovered were assigned iconic symbols like the ones traditionally used to designate the planets. By 1855 there were two dozen asteroid symbols, which often occurred in multiple variants.[4]

| Asteroid | Symbol | | Year |
|---|---|---|---|
| Ceres | 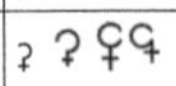 | Ceres' scythe,<br>reversed to double as the letter *C* | 1801 |
| 2 Pallas |  | Athena's (Pallas') spear | 1801 |
| 3 Juno | 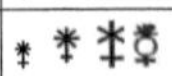 | A star mounted on a scepter,<br>for Juno, the Queen of Heaven | 1804 |
| 4 Vesta |  | The altar and sacred fire of Vesta | 1807 |
| 5 Astraea | 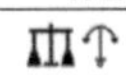 | A scale, or an inverted anchor,<br>symbols of justice | 1845 |
| 6 Hebe |  | Hebe's cup | 1847 |
| 7 Iris |  | A rainbow (*iris*) and a star | 1847 |
| 8 Flora |  | A flower (*flora*)<br>(spec. the Rose of England) | 1847 |
| 9 Metis |  | The eye of wisdom and a star | 1848 |
| 10 Hygiea |  | Hygiea's serpent and a star, or the Rod of Asclepius | 1849 |
| 11 Parthenope |  | A harp, or a fish and a star;<br>symbols of the sirens | 1850 |
| 12 Victoria |  | The laurels of victory and a star | 1850 |
| 13 Egeria |  | A shield, symbol of Egeria's protection,<br>and a star | 1850 |
| 14 Irene |  | A dove carrying an olive-branch (symbol of<br>*irene* 'peace') with a star on its head,[5] or<br>an olive branch, a flag of truce, and a star | 1851 |
| 15 Eunomia |  | A heart, symbol of good order<br>(*eunomia*), and a star | 1851 |
| 16 Psyche |  | A butterfly's wing, symbol of<br>the soul (*psyche*), and a star | 1852 |
| 17 Thetis |  | A dolphin, symbol of Thetis, and a star | 1852 |
| 18 Melpomene |  | The dagger of Melpomene, and a star | 1852 |
| 19 Fortuna |  | The wheel of fortune and a star | 1852 |
| 26 Proserpina |  | Proserpina's pomegranate | 1853 |
| 28 Bellona |  | Bellona's whip and lance[6] | 1854 |
| 29 Amphitrite |  | The shell of Amphitrite and a star | 1854 |
| 35 Leukothea |  | A lighthouse beacon,<br>symbol of Leucothea[7] | 1855 |
| 37 Fides |  | The cross of faith (*fides*)[8] | 1855 |

In 1851,[9] after the fifteenth asteroid (Eunomia) had been discovered, Johann Franz Encke made a major change in the upcoming 1854 edition of the *Berliner Astronomisches Jahrbuch* (BAJ, *Berlin Astronomical Yearbook*). He introduced a disk (circle), a traditional symbol for a star, as the generic symbol for an asteroid. The circle was then numbered in order of discovery to indicate a specific asteroid (although he assigned ① to the fifth, Astraea, while continuing to designate the first four only with their existing iconic symbols). The numbered-circle convention was quickly adopted by the astronomical community, and the next asteroid to be discovered (16 Psyche, in 1852) was the first to be designated in this manner at the time of its discovery. However, Psyche was also given an iconic symbol, as were a few other asteroids discovered over the next few years (see chart above). 20 Massalia was the first asteroid that was not assigned a symbol, and no additional iconic symbols were created after the 1855 discovery of 37 Fides.[10] That year Astraea's number was bumped up to ⑤, but Ceres through Vesta would not be listed by their numbers until the 1867 edition. The circle would become a pair of parentheses, and the parentheses sometimes omitted altogether over the next few decades, leading to the modern convention.[5]

## Discovery

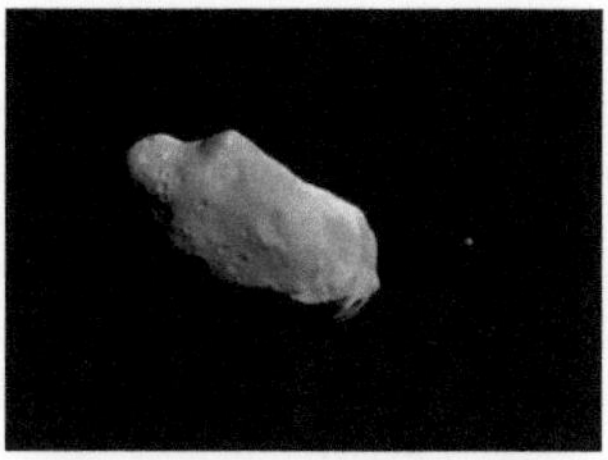

243 Ida and its moon Dactyl. Dactyl is the first satellite of an asteroid to be discovered.

The first asteroid to be discovered, Ceres, was found in 1801 by Giuseppe Piazzi, and was originally considered to be a new planet.[11] This was followed by the discovery of other similar bodies, which, with the equipment of the time, appeared to be points of light, like stars, showing little or no planetary disc, though readily distinguishable from stars due to their apparent motions. This prompted the astronomer Sir William Herschel to propose the term "asteroid", coined in Greek as αστεροειδής *asteroeidēs* 'star-like, star-shaped', from ancient Greek ἀστήρ *astēr* 'star, planet'. In the early second half of the nineteenth century, the terms "asteroid" and "planet" (not always qualified as "minor") were still used interchangeably; for example, the *Annual of Scientific Discovery for 1871* [12], page 316, reads "Professor J. Watson has been awarded by the Paris Academy of Sciences, the astronomical prize, Lalande foundation, for the discovery of eight new asteroids in one year. The planet Lydia (No. 110), discovered by M. Borelly at the Marseilles Observatory [...] M. Borelly had previously discovered two planets bearing the numbers 91 and 99 in the system of asteroids revolving between Mars and Jupiter".

### Historical methods

Asteroid discovery methods have dramatically improved over the past two centuries.

In the last years of the 18th century, Baron Franz Xaver von Zach organized a group of 24 astronomers to search the sky for the missing planet predicted at about 2.8 AU from the Sun by the Titius-Bode law, partly because of the discovery, by Sir William Herschel in 1781, of the planet Uranus at the distance predicted by the law. This task required that hand-drawn sky charts be prepared for all stars in the zodiacal band down to an agreed-upon limit of faintness. On subsequent nights, the sky would be charted again and any moving object would, hopefully, be spotted. The expected motion of the missing planet was about 30 seconds of arc per hour, readily discernible by observers.

The first object, Ceres, was not discovered by a member of the group, but rather by accident in 1801 by Giuseppe Piazzi, director of the observatory of Palermo in Sicily. He discovered a new star-like object in Taurus and followed the displacement of this object during several nights. His colleague, Carl Friedrich Gauss, used these observations to find the exact distance from this unknown object to the Earth. Gauss' calculations placed the object between the planets Mars and Jupiter. Piazzi named it after Ceres, the Roman goddess of agriculture.

Three other asteroids (2 Pallas, 3 Juno, and 4 Vesta) were discovered over the next few years, with Vesta found in 1807. After eight more years of fruitless searches, most astronomers assumed that there were no more and abandoned any further searches.

However, Karl Ludwig Hencke persisted, and began searching for more asteroids in 1830. Fifteen years later, he found 5 Astraea, the first new asteroid in 38 years. He also found 6 Hebe less than two years later. After this, other astronomers joined in the search and at least one new asteroid was discovered every year after that (except the wartime year 1945). Notable asteroid hunters of this early era were J. R. Hind, Annibale de Gasparis, Robert Luther, H. M. S. Goldschmidt, Jean Chacornac, James Ferguson, Norman Robert Pogson, E. W. Tempel, J. C. Watson, C. H. F. Peters, A. Borrelly, J. Palisa, the Henry brothers and Auguste Charlois.

In 1891, however, Max Wolf pioneered the use of astrophotography to detect asteroids, which appeared as short streaks on long-exposure photographic plates. This dramatically increased the rate of detection compared with earlier visual methods: Wolf alone discovered 248 asteroids, beginning with 323 Brucia, whereas only slightly more than 300 had been discovered up to that point. It was known that there were many more, but most astronomers did not bother with them, calling them "vermin of the skies", a phrase due to Edmund Weiss.[13] Even a century later, only a few thousand asteroids were identified, numbered and named.

### Manual methods of the 1900s and modern reporting

Until 1998, asteroids were discovered by a four-step process. First, a region of the sky was photographed by a wide-field telescope, or Astrograph. Pairs of photographs were taken, typically one hour apart. Multiple pairs could be taken over a series of days. Second, the two films or plates of the same region were viewed under a stereoscope. Any body in orbit around the Sun would move slightly between the pair of films. Under the stereoscope, the image of the body would seem to float slightly above the background of stars. Third, once a moving body was identified, its location would be measured precisely using a digitizing microscope. The location would be measured relative to known star locations.[14]

These first three steps do not constitute asteroid discovery: the observer has only found an apparition, which gets a provisional designation, made up of the year of discovery, a letter representing the half-month of discovery, and finally a letter and a number indicating the discovery's sequential number (example: 1998 $FJ_{74}$).

The last step of discovery is to send the locations and time of observations to the Minor Planet Center, where computer programs determine whether an apparition ties together earlier apparitions into a single orbit. If so, the object receives a catalogue number and the observer of the first apparition with a calculated orbit is declared the discoverer, and granted the honor of naming the object subject to the approval of the International Astronomical Union.

### Computerized methods

There is increasing interest in identifying asteroids whose orbits cross Earth's, and that could, given enough time, collide with Earth (see Earth-crosser asteroids). The three most important groups of near-Earth asteroids are the Apollos, Amors, and Atens. Various asteroid deflection strategies have been proposed, as early as the 1960s.

2004 FH is the center dot being followed by the sequence; the object that flashes by during the clip is an artificial satellite.

The near-Earth asteroid 433 Eros had been discovered as long ago as 1898, and the 1930s brought a flurry of similar objects. In order of discovery, these were: 1221 Amor, 1862 Apollo, 2101 Adonis, and finally 69230 Hermes, which approached within 0.005 AU of the Earth in 1937. Astronomers began to realize the possibilities of Earth impact.

Two events in later decades increased the alarm: the increasing acceptance of Walter Alvarez' hypothesis that an impact event resulted in the Cretaceous–Paleogene extinction, and the 1994 observation of Comet Shoemaker-Levy 9 crashing into Jupiter. The U.S. military also declassified the information that its military satellites, built to detect nuclear explosions, had detected hundreds of upper-atmosphere impacts by objects ranging from one to 10 metres across.

All these considerations helped spur the launch of highly efficient automated systems that consist of Charge-Coupled Device (CCD) cameras and computers directly connected to telescopes. Since 1998, a large majority of the asteroids have been discovered by such automated systems. A list of teams using such automated systems includes:[15]

- The Lincoln Near-Earth Asteroid Research (LINEAR) team
- The Near-Earth Asteroid Tracking (NEAT) team
- Spacewatch
- The Lowell Observatory Near-Earth-Object Search (LONEOS) team
- The Catalina Sky Survey (CSS)
- The Campo Imperatore Near-Earth Objects Survey (CINEOS) team
- The Japanese Spaceguard Association
- The Asiago-DLR Asteroid Survey (ADAS)

The LINEAR system alone has discovered 121,346 asteroids, as of March, 2011.[16] Among all the automated systems, 4711 near-Earth asteroids have been discovered[17] including over 600 more than 1 km (0.6 mi) in diameter.

## Terminology

Traditionally, small bodies orbiting the Sun were classified as asteroids, comets or meteoroids, with anything smaller than ten metres across being called a meteoroid.[18] The term "asteroid" is ill-defined. It never had a formal definition, with the broader term minor planet being preferred by the International Astronomical Union from 1853 on. In 2006, the term "small Solar System body" was introduced to cover both most minor planets and comets.[19] Other languages prefer "planetoid" (Greek for "planet-like"), and this term is occasionally used in English for the larger asteroids. The word "planetesimal" has a similar meaning, but refers specifically to the small building blocks of the planets that existed when the Solar System was forming. The term "planetule" was coined by the geologist William Daniel Conybeare to describe minor planets,[20] but is not in common use. The three largest objects in the asteroid belt, Ceres, 2 Pallas, and 4 Vesta, grew to the stage of protoplanets. Ceres has been classified as a dwarf planet, the only one in the inner Solar System.

When found, asteroids were seen as a class of objects distinct from comets, and there was no unified term for the two until "small Solar System body" was coined in 2006. The main difference between an asteroid and a comet is that a comet shows a coma due to sublimation of near surface ices by solar radiation. A few objects have ended up being dual-listed because they were first classified as minor planets but later showed evidence of cometary activity. Conversely, some (perhaps all) comets are eventually depleted of their surface volatile ices and become asteroids. A further distinction is that comets typically have more eccentric orbits than most asteroids; most "asteroids" with notably eccentric orbits are probably dormant or extinct comets.[21]

For almost two centuries, from the discovery of Ceres in 1801 until the discovery of the first centaur, 2060 Chiron, in 1977, all known asteroids spent most of their time at or within the orbit of Jupiter, though a few such as 944 Hidalgo ventured far beyond Jupiter for part of their orbit. When astronomers started finding more small bodies that permanently resided further out than Jupiter, now called centaurs, they numbered them among the traditional asteroids, though there was debate over whether they should be classified as asteroids or as a new type of object. Then, when the first trans-Neptunian object, 1992 QB1, was discovered in 1992, and especially when large numbers of similar objects started turning up, new terms were invented to sidestep the issue: Kuiper-belt object, trans-Neptunian object, scattered-disc object, and so on. These inhabit the cold outer reaches of the Solar System where ices remain solid and comet-like bodies are not expected to exhibit much cometary activity; if centaurs or trans-Neptunian objects were to venture close to the Sun, their volatile ices would sublimate, and traditional approaches would classify them as comets and not asteroids.

The innermost of these are the Kuiper-belt objects, called "objects" partly to avoid the need to classify them as asteroids or comets.[22] They are believed to be predominantly comet-like in composition, though some may be more akin to asteroids.[23] Furthermore, most do not have the highly eccentric orbits associated with comets, and the ones so far discovered are larger than traditional comet nuclei. (The much more distant Oort cloud is hypothesized to be the main reservoir of dormant comets.) Other recent observations, such as the analysis of the cometary dust collected by the Stardust probe, are increasingly blurring the distinction between comets and asteroids,[24] suggesting "a continuum between asteroids and comets" rather than a sharp dividing line.[25]

The minor planets beyond Jupiter's orbit are sometimes also called "asteroids", especially in popular presentations.[26] However, it is becoming increasingly common for the term "asteroid" to be restricted to minor planets of the inner Solar System.[22] Therefore, this article will restrict itself for the most part to the classical asteroids: objects of the asteroid belt, Jupiter trojans, and near-Earth objects.

When the IAU introduced the class small Solar System bodies in 2006 to include most objects previously classified as minor planets and comets, they created the class of dwarf planets for the largest minor planets—those that have enough mass to have become ellipsoidal under their own gravity. According to the IAU, "the term 'minor planet' may still be used, but generally the term 'Small Solar System Body' will be preferred."[27] Currently only the largest object in the asteroid belt, Ceres, at about 950 km (590 mi) across, has been placed in the dwarf planet category, although there are several large asteroids (Vesta, Pallas, and Hygiea) that may be classified as dwarf planets when their shapes are better known.[28]

## Formation

It is believed that planetesimals in the asteroid belt evolved much like the rest of the solar nebula until Jupiter neared its current mass, at which point excitation from orbital resonances with Jupiter ejected over 99% of planetesimals in the belt. Simulations and a discontinuity in spin rate and spectral properties suggest that asteroids larger than approximately 120 km (75 mi) in diameter accreted during that early era, whereas smaller bodies are fragments from collisions between asteroids during or after the Jovian disruption.[29] Ceres and Vesta grew large enough to melt and differentiate, with heavy metallic elements sinking to the core, leaving rocky minerals in the crust.[30]

In the Nice model, many Kuiper-belt objects are captured in the outer asteroid belt, at distances greater than 2.6 AU. Most were later ejected by Jupiter, but those that remained may be the D-type asteroids, and possibly include Ceres.[31]

## Distribution within the Solar System

Various dynamical groups of asteroids have been discovered orbiting in the inner Solar System. Their orbits are perturbed by the gravity of other bodies in the Solar System and by the Yarkovsky effect. Significant populations include:

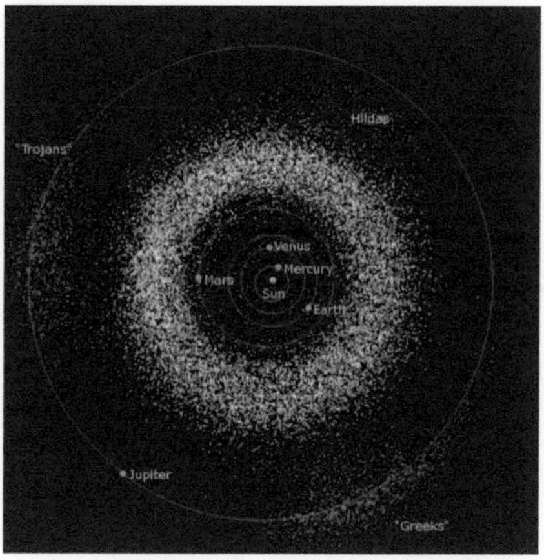

The asteroid belt (white) and the Trojan asteroids (green)

### Asteroid belt

The majority of known asteroids orbit within the asteroid belt between the orbits of Mars and Jupiter, generally in relatively low-eccentricity (i.e., not very elongated) orbits. This belt is now estimated to contain between 1.1 and 1.9 million asteroids larger than 1 km (0.6 mi) in diameter,[32] and millions of smaller ones.[33] These asteroids may be remnants of the protoplanetary disk, and in this region the accretion of planetesimals into planets during the formative period of the Solar System was prevented by large gravitational perturbations by Jupiter.

### Trojans

Trojan asteroids are a population that share an orbit with a larger planet or moon, but do not collide with it because they orbit in one of the two Lagrangian points of stability, L4 and L5, which lie 60° ahead of and behind the larger body.

The most significant population of Trojan asteroids are the Jupiter Trojans. Although fewer Jupiter Trojans have been discovered as of 2010, it is thought that they are as numerous as the asteroids in the asteroid belt.

A couple of trojans have also been found orbiting with Mars.[34]

### Near-Earth asteroids

Near-Earth asteroids, or NEAs, are asteroids that have orbits that pass close to that of Earth. Asteroids that actually cross the Earth's orbital path are known as *Earth-crossers*. As of May 2010, 7,075 near-Earth asteroids are known and the number over one kilometre in diameter is estimated to be 500–1,000.

## Characteristics

### Size distribution

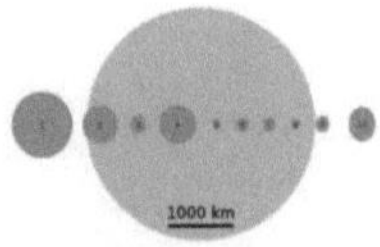


Sizes of the first ten asteroids to be discovered, compared to the Earth's Moon

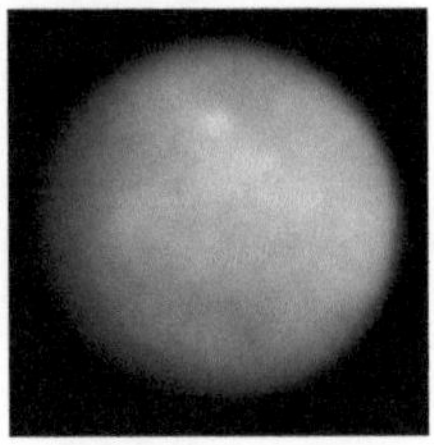

HST image of the dwarf planet Ceres

Asteroids vary greatly in size, from almost 1000 kilometres for the largest down to rocks just tens of metres across.[35] The three largest are very much like miniature planets: they are roughly spherical, have at least partly differentiated interiors,[36] and are thought to be surviving protoplanets. The vast majority, however, are much smaller and are irregularly shaped; they are thought to be either surviving planetesimals or fragments of larger bodies.

The dwarf planet Ceres is by far the largest asteroid, with a diameter of 975 km (610 mi). The next largest are 2 Pallas and 4 Vesta, both with diameters of just over 500 km (300 mi). Vesta is the only main-belt asteroid that can, on occasion, be visible to the naked eye. On some rare occasions, a near-Earth asteroid may briefly become visible without technical aid; see 99942 Apophis.

The mass of all the objects of the asteroid belt, lying between the orbits of Mars and Jupiter, is estimated to be about $2.8–3.2\times10^{21}$ kg, or about 4 percent of the mass of the Moon. Of this, Ceres comprises $0.95\times10^{21}$ kg, a third of the total.[37] Adding in the next three most massive objects, Vesta (9%), Pallas (7%), and Hygiea (3%), brings this figure up to 51%; while the three after that, 511 Davida (1.2%), 704 Interamnia (1.0%), and 52 Europa (0.9%), only add another 3% to the total mass. The number of asteroids then increases rapidly as their individual masses decrease.

The number of asteroids decreases markedly with size. Although this generally follows a power law, there are 'bumps' at 5 km and 100 km, where more asteroids than expected from a logarithmic distribution are found.[38]

## Approximate number of asteroids N larger than diameter D

| D | 100 m | 300 m | 500 m | 1 km | 3 km | 5 km | 10 km | 30 km | 50 km | 100 km | 200 km | 300 km | 500 km | 900 km |
|---|---|---|---|---|---|---|---|---|---|---|---|---|---|---|
| N | ~25,000,000 | 4,000,000 | 2,000,000 | 750,000 | 200,000 | 90,000 | 10,000 | 1,100 | 600 | 200 | 30 | 5 | 3 | 1 |

### Largest asteroids

Although their location in the asteroid belt excludes them from planet status, the four largest objects, Ceres, Vesta, Pallas, and Hygiea, are remnant protoplanets that share many characteristics common to planets, and are atypical compared to the majority of "potato"-shaped asteroids.

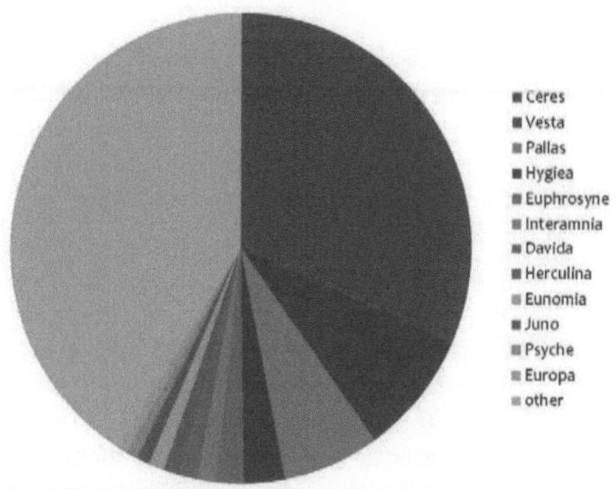


The relative masses of the twelve List of notable asteroids#Largest by masslargest asteroids known, "Recent Asteroid Mass Determinations". Maintained by Jim Baer. Last updated 2010-12-12. Access date 2011-09-02. The values of Juno and Herculina may be off by as much as 16%, and Euphrosyne by a third. The order of the lower eight may change as better data is acquired, but the values do not overlap with any known asteroid outside these twelve. compared to the remaining mass of the asteroid belt. Elena V. PitjevaPitjeva, E. V. (2005). "High-Precision Ephemerides of Planets—EPM and Determination of Some Astronomical Constants" (PDF). Solar System Research 39 (3): 184. Bibcode 2005SoSyR..39..176P. doi:10.1007/s11208-005-0033-2. .

**Attributes of protoplanetary asteroids**

| Name | Orbital radius (AU) | Orbital period (years) | Inclination to ecliptic | Orbital eccentricity | Diameter (km) | Diameter (% of Moon) | Mass ($\times 10^{18}$ kg) | Mass (% of Ceres) | Density[41] $g/cm^3$ | Rotation period (hr) | Axial tilt | Surface temperature |
|---|---|---|---|---|---|---|---|---|---|---|---|---|
| **Vesta** | 2.36 | 3.63 | 7.1° | 0.089 | 573×557×446 (mean 525) | 15% | 260 | 28% | 3.44 ± 0.12 | 5.34 | 29° | 85–270 K |
| **Ceres** | 2.77 | 4.60 | 10.6° | 0.079 | 975×975×909 (mean 952) | 28% | 940 | 100% | 2.12 ± 0.04 | 9.07 | ≈ 3° | 167 K |
| **Pallas** | 2.77 | 4.62 | 34.8° | 0.231 | 580×555×500 (mean 545) | 16% | 210 | 22% | 2.71 ± 0.11 | 7.81 | ≈ 80° | 164 K |
| **Hygiea** | 3.14 | 5.56 | 3.8° | 0.117 | 530×407×370 (mean 430) | 12% | 87 | 9% | 2.76 ± 1.2 | 27.6 | ≈ 60° | 164 K |

Ceres is the only asteroid large enough for its gravity to force it into a spheroidal shape, and so, according to the IAU's 2006 resolution on the definition of a planet, it has been classified as a dwarf planet.[42] Vesta may eventually be so classified as well. Ceres has a much higher absolute magnitude than the other asteroids, of around 3.32,[43] and may possess a surface layer of ice.[44] Like the planets, Ceres is differentiated: it has a crust, a mantle and a core.[44] Vesta, too, has a differentiated interior, though it formed inside the Solar System's frost line, and so is devoid of water;[45] its composition is mainly of basaltic rock such as olivine.[46] Pallas is unusual in that, like Uranus, it rotates on its side, with one pole regularly facing the Sun and the other facing away.[47] Its composition is similar to that of Ceres: high in carbon and silicon, and perhaps partially differentiated.[48] Hygiea is a carbonaceous asteroid and, unlike the other largest asteroids, lies relatively close to the plane of the ecliptic.[49]

## Rotation

Measurements of the rotation rates of large asteroids in the asteroid belt show that there is an upper limit. No asteroid with a diameter larger than 100 meters has a rotation period smaller than 2.2 hours. For asteroids rotating faster than approximately this rate, the inertia at the surface is greater than the gravitational force, so any loose surface material would be flung out. However, a solid object should be able to rotate much more rapidly. This suggests that most asteroids with a diameter over 100 meters are rubble piles formed through accumulation of debris after collisions between asteroids.[50]

## Composition

The physical composition of asteroids is varied and in most cases poorly understood. Ceres appears to be composed of a rocky core covered by an icy mantle, where Vesta is thought to have a nickel-iron core, olivine mantle, and basaltic crust.[51] 10 Hygiea, however, which appears to have a uniformly primitive composition of carbonaceous chondrite, is thought to be the largest undifferentiated asteroid. Most of the smaller asteroids are thought to be piles of rubble held together loosely by gravity, though the largest are probably solid. Some asteroids have moons or are co-orbiting binaries: Rubble piles, moons, binaries, and scattered asteroid families are believed to be the results of collisions that disrupted a parent asteroid.

Asteroids contain traces of amino acids and other organic compounds, and some speculate that asteroid impacts may have seeded the early Earth with the chemicals necessary to initiate life, or may have even brought life itself to Earth. (See also panspermia.)[52] In August 2011, a report, based on NASA studies with meteorites found on Earth, was published suggesting DNA and RNA components (adenine, guanine and related organic molecules) may have been formed on asteroids and comets in outer space.[53] [54] [55]

Only one asteroid, 4 Vesta, which has a reflective surface, is normally visible to the naked eye, and this only in very dark skies when it is favorably positioned. Rarely, small asteroids passing close to Earth may be naked-eye visible for a short time.[56]

Composition is calculated from three primary sources: albedo, surface spectrum, and density. The last can only be determined accurately by observing the orbits of moons the asteroid might have. So far, every asteroid with moons has turned out to be a rubble pile, a loose conglomeration of rock and metal that may be half empty space by volume. The investigated asteroids are as large as 280 km in diameter, and include 121 Hermione (268×186×183 km), and 87 Sylvia (384×262×232 km). Only half a dozen asteroids are larger than 87 Sylvia, though none of them have moons; however, some smaller asteroids are thought to be more massive, suggesting they may not have been disrupted, and indeed 511 Davida, the same size as Sylvia to within measurement error, is estimated to be two and a half times as massive, though this is highly uncertain. The fact that such large asteroids as Sylvia can be rubble piles, presumably due to disruptive impacts, has important consequences for the formation of the Solar system: Computer simulations of collisions involving solid bodies show them destroying each other as often as merging, but colliding rubble piles are more likely to merge. This means that the cores of the planets could have formed relatively quickly.[57]

### Surface features

253 Mathilde, a C-type asteroid measuring about 50 kilometres (31.07 mi) across, covered in craters half that size. Photograph taken in 1997 by the NEAR Shoemaker probe.

Most asteroids outside the big four (Ceres, Pallas, Vesta, and Hygiea) are likely to be broadly similar in appearance, if irregular in shape. 50-km 253 Mathilde (shown at right) is a rubble pile saturated with craters with diameters the size of the asteroid's radius, and Earth-based observations of 300-km 511 Davida, one of the largest asteroids after the big four, reveal a similarly angular profile, suggesting it is also saturated with radius-size craters.[58] Medium-sized asteroids such as Mathilde and 243 Ida that have been observed up close also reveal a deep regolith covering the surface. Of the big four, Pallas and Hygiea are practically unknown. Vesta has compression fractures encircling a radius-size crater at its south pole but is otherwise a spheroid. Ceres seems quite different in the glimpses Hubble has provided, with surface features that are unlikely to be due to simple craters and impact basins, but details will not be known until *Dawn* arrives in 2015.

## Classification

Asteroids are commonly classified according to two criteria: the characteristics of their orbits, and features of their reflectance spectrum.

### Orbital classification

Many asteroids have been placed in groups and families based on their orbital characteristics. Apart from the broadest divisions, it is customary to name a group of asteroids after the first member of that group to be discovered. Groups are relatively loose dynamical associations, whereas families are tighter and result from the catastrophic break-up of a large parent asteroid sometime in the past.[59] Families have only been recognized within the asteroid belt. They were first recognised by Kiyotsugu Hirayama in 1918 and are often called Hirayama families in his honor.

About 30% to 35% of the bodies in the asteroid belt belong to dynamical families each thought to have a common origin in a past collision between asteroids. A family has also been associated with the plutoid dwarf planet Haumea.

#### Quasi-satellites and horseshoe objects

Some asteroids have unusual horseshoe orbits that are co-orbital with the Earth or some other planet. Examples are 3753 Cruithne and 2002 $AA_{29}$. The first instance of this type of orbital arrangement was discovered between Saturn's moons Epimetheus and Janus.

Sometimes these horseshoe objects temporarily become quasi-satellites for a few decades or a few hundred years, before returning to their earlier status. Both Earth and Venus are known to have quasi-satellites.

Such objects, if associated with Earth or Venus or even hypothetically Mercury, are a special class of Aten asteroids. However, such objects could be associated with outer planets as well.

### Spectral classification

This picture of 433 Eros shows the view looking from one end of the asteroid across the gouge on its underside and toward the opposite end. Features as small as 35 m (115 ft) across can be seen.

In 1975, an asteroid taxonomic system based on colour, albedo, and spectral shape was developed by Clark R. Chapman, David Morrison, and Ben Zellner.[60] These properties are thought to correspond to the composition of the asteroid's surface material. The original classification system had three categories: C-types for dark carbonaceous objects (75% of known asteroids), S-types for stony (silicaceous) objects (17% of known asteroids) and U for those that did not fit into either C or S. This classification has since been expanded to include many other asteroid types. The number of types continues to grow as more asteroids are studied.

The two most widely used taxonomies now used are the Tholen classification and SMASS classification. The former was proposed in 1984 by David J. Tholen, and was based on data collected from an eight-color asteroid survey performed in the 1980s. This resulted in 14 asteroid categories.[61] In 2002, the Small Main-Belt Asteroid Spectroscopic Survey resulted in a modified version of the Tholen taxonomy with 24 different types. Both systems have three broad categories of C, S, and X asteroids, where X consists of mostly metallic asteroids, such as the M-type. There are also several smaller classes.[62]

Note that the proportion of known asteroids falling into the various spectral types does not necessarily reflect the proportion of all asteroids that are of that type; some types are easier to detect than others, biasing the totals.

#### Problems

Originally, spectral designations were based on inferences of an asteroid's composition.[63] However, the correspondence between spectral class and composition is not always very good, and a variety of classifications are in use. This has led to significant confusion. While asteroids of different spectral classifications are likely to be composed of different materials, there are no assurances that asteroids within the same taxonomic class are composed of similar materials.

At present, the spectral classification based on several coarse resolution spectroscopic surveys in the 1990s is still the standard. Scientists cannot agree on a better taxonomic system, largely due to the difficulty of obtaining detailed measurements consistently for a large sample of asteroids (e.g. finer resolution spectra, or non-spectral data such as densities would be very useful).

## Exploration

Until the age of space travel, objects in the asteroid belt were merely pinpricks of light in even the largest telescopes and their shapes and terrain remained a mystery. The best modern ground-based telescopes and the Earth-orbiting Hubble Space Telescope can resolve a small amount of detail on the surfaces of the largest asteroids, but even these mostly remain little more than fuzzy blobs. Limited information about the shapes and compositions of asteroids can be inferred from their light curves (their variation in brightness as they rotate) and their spectral properties, and asteroid sizes can be estimated by timing the lengths of star occulations (when an asteroid passes directly in front of a star). Radar imaging can yield good information about asteroid shapes and orbital and rotational parameters, especially for near-Earth asteroids. In terms of delta-v and propellant requirements, NEOs are more easily accessible than the Moon.[64]

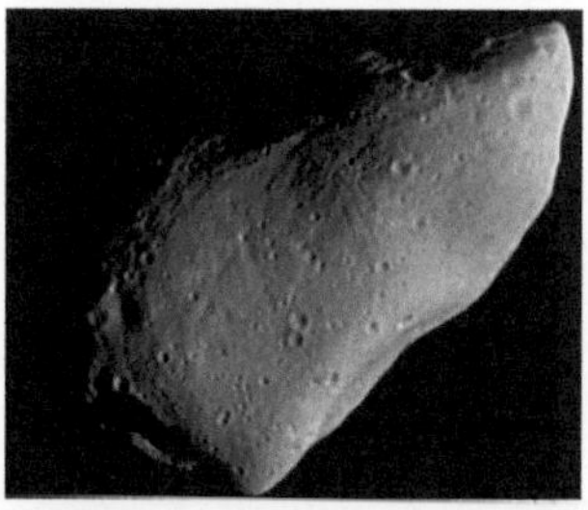

951 Gaspra is the first asteroid to be imaged in close-up.

Vesta, imaged by the Dawn spacecraft

The first close-up photographs of asteroid-like objects were taken in 1971 when the Mariner 9 probe imaged Phobos and Deimos, the two small moons of Mars, which are probably captured asteroids. These images revealed the irregular, potato-like shapes of most asteroids, as did later images from the Voyager probes of the small moons of the gas giants.

The first true asteroid to be photographed in close-up was 951 Gaspra in 1991, followed in 1993 by 243 Ida and its moon Dactyl, all of which were imaged by the Galileo probe en route to Jupiter.

The first dedicated asteroid probe was NEAR Shoemaker, which photographed 253 Mathilde in 1997, before entering into orbit around 433 Eros, finally landing on its surface in 2001.

Other asteroids briefly visited by spacecraft en route to other destinations include 9969 Braille (by Deep Space 1 in 1999), and 5535 Annefrank (by Stardust in 2002).

In September 2005, the Japanese Hayabusa probe started studying 25143 Itokawa in detail and was plagued with difficulties, but returned samples of its surface to earth on June 13, 2010.

The European Rosetta probe (launched in 2004) flew by 2867 Šteins in 2008 and 21 Lutetia, the second-largest asteroid visited to date, in 2010.

In September 2007, NASA launched the Dawn Mission, which orbited the protoplanet 4 Vesta from July 2011 to September 2012, and is planned to orbit 1 Ceres in 2015. 4 Vesta is the largest asteroid visited to date.

On 13 December 2012, China's lunar orbiter Chang'e 2 flew within 2 miles (3 km) of the asteroid 4179 Toutatis on an extended mission.

The Japan Aerospace Exploration Agency (JAXA) plans to launch around 2015 the improved Hayabusa 2 space probe and to return asteroid samples by 2020. Current target for the mission is the C-type asteroid (162173) 1999 $JU_3$.

In May 2011, NASA announced the OSIRIS-REx sample return mission to asteroid 1999 RQ36, and is expected to launch in 2016.

It has been suggested that asteroids might be used as a source of materials that may be rare or exhausted on earth (asteroid mining), or materials for constructing space habitats (see Colonization of the asteroids). Materials that are heavy and expensive to launch from earth may someday be mined from asteroids and used for space manufacturing and construction.

## Fiction

Asteroids and the asteroid belt are a staple of science fiction stories. Asteroids play several potential roles in science fiction: as places human beings might colonize, resources for extracting minerals, hazards encountered by spaceships traveling between two other points, and as a threat to life on Earth by potential impact.

## See also

- BOOTES (Burst Observer and Optical Transient Exploring System)
- Asteroid deflection strategies
- Meanings of asteroid names
- Pronunciation of asteroid names
- List of asteroids named after people
- List of asteroids named after places
- Category:Asteroid groups and families
- Category:Binary asteroids
- Category:Asteroids
- List of minor planets
- List of notable asteroids
- Centaur (minor planet)
- Dwarf planet
- Impact event
- Lost asteroid
- Mesoplanet
- Minor planet
- Minor Planet Center
- Marco Polo (spacecraft)
- Near-Earth object
- Orion asteroid mission
- Pioneer 10 space probe
- Rosetta probe

## Notes

[1] "Asteroids" (http://ssd.jpl.nasa.gov/?asteroids). NASA – Jet Propulsion Laboratory. . Retrieved 13 September2010.

[2] Asimov, Isaac, and Dole, Stephen H. *Planets for Man* (New York: Random House, 1964), p.43.

[3] "What Are Asteroids And Comets?" (http://neo.jpl.nasa.gov/faq/#ast). *Near Earth Object Program FAQ*. NASA. Archived (http://web.archive.org/web/20100909210213/http://neo.jpl.nasa.gov/faq/) from the original on 9 September 2010. . Retrieved 2010-09-13.

[4] Gould, B. A. (1852). "On the Symbolic Notation of the Asteroids". *Astronomical Journal* **2**: 80. Bibcode 1852AJ......2...80G. doi:10.1086/100212.

[5] Hilton, James L. (2001-09-17). "When Did the Asteroids Become Minor Planets" (http://aa.usno.navy.mil/faq/docs/minorplanets.php). . Retrieved 2006-03-26.

[6] Encke, J. F. (1854). "Beobachtung der Bellona, nebst Nachrichten über die Bilker Sternwarte". *Astronomische Nachrichten* **38** (9): 143. doi:10.1002/asna.18540380907.

[7] Rümker, G. (1855). "Name und Zeichen des von Herrn R. Luther zu Bilk am 19. April entdeckten Planeten". *Astronomische Nachrichten* **40** (24): 373. doi:10.1002/asna.18550402405.

[8] Luther, R. (1856). "Schreiben des Herrn Dr. R. Luther, Directors der Sternwarte zu Bilk, an den Herausgeber". *Astronomische Nachrichten* **42** (7): 107. Bibcode 1855AN.....42..107L. doi:10.1002/asna.18550420705.

[9] "When did the asteroids become minor planets?" (http://www.usno.navy.mil/USNO/astronomical-applications/astronomical-information-center/minor-planets). Naval Meteorology and Oceanography Command. . Retrieved 2011-11-06.

[10] Except for Pluto and, in the astrological community, for a few outer bodies such as 2060 Chiron

[11] Ceres is the largest asteroid and is now classified as a dwarf planet. All other asteroids are now classified as small Solar System bodies along with comets, centaurs, and the smaller trans-Neptunian objects.

[12] http://books.google.ca/books?id=NAMAAAAAMAAJ&pg=PA316&dq=%22planets%22+asteroids

[13] Seares, Frederick H. (1930). "Address of the Retiring President of the Society in Awarding the Bruce Medal to Professor Max Wolf". *Publ. Astr. Soc. Pacific* **42**: 5–22. Bibcode 1930PASP...42....5S. doi:10.1086/123986+(Dead+links).

[14] Chapman, Mary G. (May 17, 1992). "Carolyn Shoemaker, Planetary Astronomer and Most Successful 'Comet Hunter' To Date" (http://astrogeology.usgs.gov/About/People/CarolynShoemaker). USGS. . Retrieved 2008-04-15.

[15] Yeomans, Don. "Near Earth Object Search Programs" (http://neo.jpl.nasa.gov/programs/). NASA. Archived (http://web.archive.org/web/20080424093951/http://neo.jpl.nasa.gov/programs/) from the original on 24 April 2008. . Retrieved 2008-04-15.

[16] "Minor Planet Discover Sites" (http://www.minorplanetcenter.org/iau/lists/MPDiscSites.html). Archived (http://web.archive.org/web/20100830074958/http://www.minorplanetcenter.org/iau/lists/MPDiscSites.html) from the original on 30 August 2010. . Retrieved 2010-08-24.

[17] "Unusual Minor Planets" (http://www.minorplanetcenter.org/iau/lists/Unusual.html). Archived (http://web.archive.org/web/20100827234833/http://www.minorplanetcenter.org/iau/lists/Unusual.html) from the original on 27 August 2010. . Retrieved 2010-08-24.

[18] Beech, M. (September 1995). "On the Definition of the Term Meteoroid". *Quarterly Journal of the Royal Astronomical Society* **36** (3): 281–284. Bibcode 1995QJRAS..36..281B.

[19] The definition of "small Solar System bodies" says that they "include most of the Solar System asteroids, most trans-Neptunian objects, comets, and other small bodies". *The Final IAU Resolution on the definition of "planet" ready for voting* (http://www.iau2006.org/mirror/www.iau.org/iau0602/index.html) (IAU)

[20] "English Dictionary – Browsing Page P-44" (http://www.hyperdictionary.com/dict-e/p-44.html). HyperDictionary.com. . Retrieved 2008-04-15.

[21] Weissman, Paul R., William F. Bottke, Jr., and Harold F. Levinson. "Evolution of Comets into Asteroids." *Southwest Research Institute, Planetary Science Directorate.* 2002. Web Retrieved 3 Aug. 2010 (http://www.boulder.swri.edu/~hal/PDF/asteroids3.pdf)

[22] "Are Kuiper Belt Objects asteroids?" (http://curious.astro.cornell.edu/question.php?number=601), "Ask an astronomer", Cornell University

[23] "Asteroids and Comets" (http://rst.gsfc.nasa.gov/Sect19/Sect19_22.html), NASA website

[24] "Comet Dust Seems More Asteroidy" (http://www.sciam.com/podcast/episode.cfm?id=ADD0878B-D6C3-3B70-7B5BC373545BB82D) *Scientific American*, January 25, 2008

[25] "Comet samples are surprisingly asteroid-like" (http://space.newscientist.com/channel/solar-system/comets-asteroids/dn13224-comet-samples-are-surprisingly-asteroidlike.html), *New Scientist*, 24 January 2008

[26] For instance, a joint NASA-JPL public-outreach website states:

> "We include Trojans (bodies captured in Jupiter's 4th and 5th Lagrange points), Centaurs (bodies in orbit between Jupiter and Neptune), and trans-Neptunian objects (orbiting beyond Neptune) in our definition of "asteroid" as used on this site, even though they may more correctly be called "minor planets" instead of asteroids."

<http://ssd.jpl.nasa.gov/?asteroids>

[27] Questions and Answers on Planets (http://www.iau.org/public_press/news/release/iau0603/questions_answers/), IAU

[28] "Three new planets may join solar system" (http://space.newscientist.com/article.ns?id=dn9761&feedId=online-news_rss20F53), *New Scientist*, 16 August 2006

[29] Bottke, Durda; Nesvorny, Jedicke; Morbidelli, Vokrouhlicky; Levison (2005). "The fossilized size distribution of the main asteroid belt" (http://astro.mff.cuni.cz/davok/papers/fossil05.pdf). *Icarus* **175**: 111. Bibcode 2005Icar..175..111B. doi:10.1016/j.icarus.2004.10.026. .

[30] Kerrod, Robin (2000). *Asteroids, Comets, and Meteors*. Lerner Publications Co.. ISBN 0-585-31763-1.

[31] William B. McKinnon, 2008, "On The Possibility Of Large KBOs Being Injected Into The Outer Asteroid Belt". (http://adsabs.harvard.edu/abs/2008DPS....40.3803M) *American Astronomical Society,* DPS meeting #40, #38.03

[32] Tedesco, Edward; Metcalfe, Leo (April 4, 2002). "New study reveals twice as many asteroids as previously believed" (http://www.spaceref.com/news/viewpr.html?pid=7925) (Press release). European Space Agency. . Retrieved 2008-02-21.

[33] World Book at NASA (http://www.nasa.gov/worldbook/asteroid_worldbook.html)

[34] Neptune also has a few known trojans, and these are thought to actually be much more numerous than the Jovian trojans. However, they are often included in the trans-Neptunian population rather than counted with the asteroids.

[35] Below 10 metres, these rocks are by convention considered to be meteoroids.

[36] Schmidt, B.; Russell, C. T.; Bauer, J. M.; Li, J.; McFadden, L. A.; Mutchler, M.; Parker, J. W.; Rivkin, A. S.; Stern, S. A.; Thomas, P. C. (2007). "Hubble Space Telescope Observations of 2 Pallas". *American Astronomical Society, DPS meeting #39* **39**: 485. Bibcode 2007DPS....39.3519S.

[37] Pitjeva, E. V. (2004). "Estimations of masses of the largest asteroids and the main asteroid belt from ranging to planets, Mars orbiters and landers" (http://adsabs.harvard.edu/abs/2004cosp.meet.2014P). *35th COSPAR Scientific Assembly. Held 18–25 July 2004, in Paris, France*. pp. 2014. .

[38] Davis 2002, "Asteroids III", cited by Željko Ivezić (http://www.astro.washington.edu/users/ivezic/Astr598/lecture4.pdf)

[39] "Recent Asteroid Mass Determinations" (http://home.earthlink.net/~jimbaer1/astmass.txt). Maintained by Jim Baer. Last updated 2010-12-12. Access date 2011-09-02. The values of Juno and Herculina may be off by as much as 16%, and Euphrosyne by a third. The order of the lower eight may change as better data is acquired, but the values do not overlap with any known asteroid outside these twelve.

[40] Pitjeva, E. V. (2005). "High-Precision Ephemerides of Planets—EPM and Determination of Some Astronomical Constants" (http://iau-comm4.jpl.nasa.gov/EPM2004.pdf) (PDF). *Solar System Research* **39** (3): 184. Bibcode 2005SoSyR..39..176P. doi:10.1007/s11208-005-0033-2. .

[41] "Asteroid Density, Porosity, and Structure" (http://www.lpi.usra.edu/books/AsteroidsIII/pdf/3022.pdf). lpi.usra.edu. . Retrieved 2013-01-03.

[42] "The Final IAU Resolution on the Definition of "Planet" Ready for Voting" (http://www.iau.org/public_press/news/detail/iau0602/). IAU. August 24, 2006. . Retrieved 2007-03-02.

[43] Parker, J. W.; Stern, S. A.; Thomas, P. C.; Festou, M. C.; Merline, W. J.; Young, E. F.; Binzel, R. P.; and Lebofsky, L. A. (2002). "Analysis of the First Disk-resolved Images of Ceres from Ultraviolet Observations with the Hubble Space Telescope". *The Astronomical Journal* **123** (1): 549–557. arXiv:astro-ph/0110258. Bibcode 2002AJ....123..549P. doi:10.1086/338093.

[44] "Asteroid 1 Ceres" (http://www.planetary.org/explore/topics/asteroids_and_comets/ceres.html). *The Planetary Society*. Archived (http://web.archive.org/web/20070929092440/http://planetary.org/explore/topics/asteroids_and_comets/ceres.html) from the original

on 29 September 2007. . Retrieved 2007-10-20.

[45] "Key Stages in the Evolution of the Asteroid Vesta" (http://hubblesite.org/newscenter/newsdesk/archive/releases/1995/20/image/c). *Hubble Space Telescope news release*. 1995. Archived (http://web.archive.org/web/20070930203613/http://hubblesite.org/newscenter/newsdesk/archive/releases/1995/20/image/c) from the original on 30 September 2007. . Retrieved 2007-10-20. Russel, C. T.; *et al.* (2007). "Dawn mission and operations" (http://journals.cambridge.org/action/displayAbstract?fromPage=online&aid=414750). *NASA/JPL*. . Retrieved 2007-10-20.

[46] Burbine, T. H. (July 1994). "Where are the olivine asteroids in the main belt?" (http://articles.adsabs.harvard.edu/cgi-bin/nph-iarticle_query?1994Metic..29..453B&data_type=PDF_HIGH&whole_paper=YES&type=PRINTER&filetype=.pdf). *Meteoritics* **29** (4): 453. Bibcode 1994Metic..29..453B. .

[47] Torppa, J.; *et al.* (1996). "Shapes and rotational properties of thirty asteroids from photometric data". *Icarus* **164** (2): 346–383. Bibcode 2003Icar..164..346T. doi:10.1016/S0019-1035(03)00146-5.

[48] Larson, H. P.; Feierberg, M. A.; and Lebofsky, L. A. (1983). "The composition of asteroid 2 Pallas and its relation to primitive meteorites" (http://adsabs.harvard.edu/abs/1983Icar...56..398L). . Retrieved 2007-10-20.

[49] Barucci, M. A.; *et al.* (2002). "10 Hygiea: ISO Infrared Observations" (http://www.lesia.obspm.fr/~crovisier/biblio/preprint/bar02_icarus.pdf) (PDF). Archived (http://web.archive.org/web/20071128200223/http://www.lesia.obspm.fr/~crovisier/biblio/preprint/bar02_icarus.pdf) from the original on 28 November 2007. . Retrieved 2007-10-21. "Ceres the Planet" (http://www.orbitsimulator.com/gravity/articles/ceres.html). *orbitsimulator.com*. Archived (http://web.archive.org/web/20071011154140/http://orbitsimulator.com/gravity/articles/ceres.html) from the original on 11 October 2007. . Retrieved 2007-10-20.

[50] Rossi, Alessandro (2004-05-20). "The mysteries of the asteroid rotation day" (http://spaceguard.iasf-roma.inaf.it/tumblingstone/issues/current/eng/ast-day.htm). The Spaceguard Foundation. . Retrieved 2007-04-09.

[51] HubbleSite - NewsCenter - Asteroid or Mini-Planet? Hubble Maps the Ancient Surface of Vesta (04/19/1995) - Release Images (http://hubblesite.org/newscenter/archive/releases/1995/20/image/)

[52] Life is Sweet: Sugar-Packing Asteroids May Have Seeded Life on Earth (http://www.space.com/scienceastronomy/planetearth/meteor_sugar_011219.html) Archived (http://web.archive.org/20020124092631/http://web.archive.org/web/20020124092631/http://www.space.com/scienceastronomy/planetearth/meteor_sugar_011219.html) January 24, 2002 at the Wayback Machine, Space.com, 19 December 2001

[53] Callahan, M.P.; Smith, K.E.; Cleaves, H.J.; Ruzica, J.; Stern, J.C.; Glavin, D.P.; House, C.H.; Dworkin, J.P. (11 August 2011). "Carbonaceous meteorites contain a wide range of extraterrestrial nucleobases" (http://www.pnas.org/content/early/2011/08/10/1106493108). PNAS. doi:10.1073/pnas.1106493108. . Retrieved 2011-08-15.

[54] Steigerwald, John (8 August 2011). "NASA Researchers: DNA Building Blocks Can Be Made in Space" (http://www.nasa.gov/topics/solarsystem/features/dna-meteorites.html). NASA. . Retrieved 2011-08-10.

[55] ScienceDaily Staff (9 August 2011). "DNA Building Blocks Can Be Made in Space, NASA Evidence Suggests" (http://www.sciencedaily.com/releases/2011/08/110808220659.htm). ScienceDaily. . Retrieved 2011-08-09.

[56] Closest Flyby of Large Asteroid to be Naked-Eye Visible (http://www.space.com/spacewatch/050204_2004_mn4.html), Space.com, 4 February 2005

[57] Marchis, Descamps, et al. *Icarus*, Feb. 2011

[58] A.R. Conrad *et al.* 2007. " Direct measurement of the size, shape, and pole of 511 Davida with Keck AO in a single night (http://www2.keck.hawaii.edu/inst/people/conrad/research/pub/d511.pdf)", *Icarus,* doi:10.1016/j.icarus.2007.05.004

[59] Zappalà, V. (1995). "Asteroid families: Search of a 12,487-asteroid sample using two different clustering techniques". *Icarus* **116** (2): 291–314. Bibcode 1995Icar..116..291Z. doi:10.1006/icar.1995.1127.

[60] Chapman, C. R. (1975). "Surface properties of asteroids: A synthesis of polarimetry, radiometry, and spectrophotometry". *Icarus* **25** (1): 104–130. Bibcode 1975Icar...25..104C. doi:10.1016/0019-1035(75)90191-8.

[61] Tholen, D. J. (March 8–11, 1988). "Asteroid taxonomic classifications" (http://adsabs.harvard.edu/abs/1989aste.conf.1139T). *Asteroids II; Proceedings of the Conference*. Tucson, AZ: University of Arizona Press. pp. 1139–1150. . Retrieved 2008-04-14.

[62] Bus, S. J. (2002). "Phase II of the Small Main-belt Asteroid Spectroscopy Survey: A feature-based taxonomy". *Icarus* **158** (1): 146. Bibcode 2002Icar..158..146B. doi:10.1006/icar.2002.6856.

[63] McSween Jr., Harry Y. (1999). *Meteorites and their Parent Planets* (2nd ed.). Oxford University Press. ISBN 0-521-58751-4.

[64] A Piloted Orion Flight to a Near-Earth Object: A Feasibility Study (http://ti.arc.nasa.gov/m/project/neo/pdf/NEO_feasibility.pdf)

## References

## External links

- Rocks from the Main Belt asteroids (http://rocksfromspace.open.ac.uk/asteroid_belt_detail.htm)
- Alphabetical list of minor planet names (ASCII) (http://www.minorplanetcenter.org/iau/lists/MPNames.html) (Minor Planet Center)
- Near Earth Asteroid Tracking (NEAT) (http://neat.jpl.nasa.gov/)
- Asteroids Page (http://solarsystem.nasa.gov/planets/profile.cfm?Object=Asteroids) at NASA's Solar System Exploration (http://solarsystem.nasa.gov/)
- Asteroid Simulator with Moon and Earth (http://www.colorado.edu/physics/2000/applets/satellites.html)
- Alphabetical and numerical lists of minor planet names (Unicode) (http://www.ipa.nw.ru/PAGE/DEPFUND/LSBSS/englenam.htm) (Institute of Applied Astronomy)
- Future Asteroid Interception Research (http://www.fair-society.org/)
- Near Earth Objects Dynamic Site (http://newton.dm.unipi.it/cgi-bin/neodys/neoibo)
- Asteroids Dynamic Site (http://hamilton.dm.unipi.it/astdys/) Up-to-date osculating orbital elements and proper orbital elements University of Pisa, Italy.
- JPL small bodies database (http://ssd.jpl.nasa.gov/?sb_elem) Current down-loadable ASCII table of orbit data and absolute mags H for over 200000 asteroids, sorted by number. Caltech/JPL.
- Asteroid naming statistics (http://quasar.ipa.nw.ru/PAGE/DEPFUND/LSBSS/statmpn.htm)
- Spaceguard UK (http://www.spaceguarduk.com/)
- Committee on Small Body Nomenclature (http://www.ss.astro.umd.edu/IAU/csbn/)
- List of minor planet orbital groupings and families from ProjectPluto (http://www.projectpluto.com/mp_group.htm)
- Cunningham, Clifford, "Introduction to Asteroids: The Next Frontier", ISBN 0-943396-16-6
- James L. Hilton: When Did the Asteroids Become Minor Planets? (http://aa.usno.navy.mil/faq/docs/minorplanets.php)
- Kirkwood, Daniel; *Relations between the Motions of some of the Minor Planets* (1874). (http://adsabs.harvard.edu/cgi-bin/nph-bib_query?bibcode=1874MNRAS..35...61K&db_key=AST&high=40daf3f6f901929)
- Schmadel, L.D. (2003). *Dictionary of Minor Planet Names.* 5th ed. IAU/Springer-Verlag: Heidelberg.
- Asteroid articles in Planetary Science Research Discoveries (http://www.psrd.hawaii.edu/Archive/Archive-Asteroids.html)
- Catalogue of the Solar System Small Bodies Orbital Evolution (http://smallbodies.ru/en/)
- TECA Table of next close approaches to the Earth (http://www.brera.mi.astro.it/sormano/teca.html)
- SAEL Small Asteroids Encounter List (http://www.brera.mi.astro.it/sormano/sael.html)
- MBPL Minor Body Priority List (http://www.brera.mi.astro.it/sormano/mbpl.html)
- PCEL Planetary Close Encounter List (http://www.brera.mi.astro.it/sormano/pcel.html)

# Minor_planet

A **minor planet** is an astronomical object in direct orbit around the Sun that is neither a dominant planet nor originally classified as a comet. Minor planets can be dwarf planets, asteroids, trojans, centaurs, Kuiper belt objects, and other trans-Neptunian objects.[1] The first minor planet discovered was Ceres in 1801 (although from the time of its discovery until 1851 it was considered to be a planet). The orbits of more than 570,000 objects have been archived at the Minor Planet Center.[2]

The term "minor planet" has been used since the 19th century to describe these objects.[3] The term **planetoid** has also been used, especially for larger (planetary) objects.[4] Historically, the terms *asteroid*, *minor planet*, and *planetoid* have been more or less synonymous,[4] [5] but the issue has been complicated by the discovery of numerous minor planets beyond the orbit of Jupiter and especially Neptune that are not universally considered asteroids.[5] Minor planets seen outgassing may receive a dual classification as a comet.

Before 2006 the International Astronomical Union had officially used the term *minor planet*. During its 2006 meeting, the Union reclassified minor planets and comets into dwarf planets and small Solar System bodies.[6] Objects are called dwarf planets if their self-gravity is sufficient to achieve hydrostatic equilibrium, that is, an ellipsoidal shape, with all other minor planets and comets called "small Solar System bodies".[6] The IAU states: "the term 'minor planet' may still be used, but generally the term 'small solar system body' will be preferred."[7] However, for purposes of numbering and naming, the traditional distinction between minor planet and comet is still followed.

## Populations

Hundreds of thousands of minor planets have been discovered within the Solar System, with the 2009 rate of discovery at over 3,000 per month. Of the more than 535,000 registered minor planets, 251,651 have orbits known well enough to be assigned permanent official numbers.[8] [2] Of these, 16,154 have official names[2] . As of September 2010, the lowest-numbered unnamed minor planet is (3708) 1974 $FV_1$;[9] but there are also some named minor planets above number 240,000.[10]

There are various broad minor-planet populations:

- Asteroids; traditionally, most have been bodies in the inner Solar System.[5]
  - Main-belt asteroids, those following roughly circular orbits between Mars and Jupiter. These are the original and best-known group of asteroids or minor planets.
  - Near-Earth asteroids, those whose orbits take them inside the orbit of Mars. Further subclassification of these, based on orbital distance, is used:[11]
    - Aten asteroids, those that have semi-major axes of less than one Earth orbit. Those Aten asteroids that have their aphelion within Earth's orbit are known as Apohele asteroids;
    - Amor asteroids are those near-Earth asteroids that approach the orbit of the Earth from beyond, but do not cross it. Amor asteroids are further subdivided into four subgroups, depending on where their semimajor axis falls between Earth's orbit and the asteroid belt;
    - Apollo asteroids are those asteroids with a semimajor axis greater than Earth's, while having a perihelion distance of 1.017 AU or less. Like Aten asteroids, Apollo asteroids are Earth-crossers.
  - Earth trojans, asteroids sharing Earth's orbit and gravitationally locked to it. As of 2011, the only one known is 2010 $TK_7$.[12]
  - Mars trojans, asteroids sharing Mars's orbit and gravitationally locked to it. As of 2007, eight such asteroids are known.[13]
  - Jupiter trojans, asteroids sharing Jupiter's orbit and gravitationally locked to it. Numerically they are estimated to equal the main-belt asteroids.
- Distant minor planets; an umbrella term for minor planets in the outer Solar System.

- Centaurs, bodies in the outer Solar System between Jupiter and Neptune. They have unstable orbits due to the gravitational influence of the giant planets, and therefore must have come from elsewhere, probably outside Neptune.[14]
- Neptune trojans, bodies sharing Neptune's orbit and gravitationally locked to it. Although only a handful are known, there is evidence that Neptune trojans are more numerous than either the asteroids in the asteroid belt or the Jupiter trojans.[15]
- Trans-Neptunian objects, bodies at or beyond the orbit of Neptune, the outermost planet.
  - The Kuiper belt, objects inside an apparent population drop-off approximately 55 AU from the Sun.
    - Classical Kuiper belt objects, also known as cubewanos, are in primordial, relatively circular orbits that are not in resonance with Neptune.
    - Plutinos, bodies like Pluto that are in a 2:3 resonance with Neptune, and other resonant trans-Neptunian objects. This is a rather large group.
  - Scattered disc, objects with aphelia outside the Kuiper belt. These are thought to have been scattered by Neptune.
  - Detached objects such as Sedna, with both aphelia and perihelia outside the Kuiper belt.
  - The Oort Cloud, a hypothetical population thought to be the source of long-period comets that may extend out to 50,000 AU from the Sun.

# Naming

A newly discovered minor planet is given a provisional designation (such as 2002 $AT_4$) consisting of the year of discovery and an alphanumeric code indicating the half-month of discovery and the sequence within that half-month. Once an asteroid's orbit has been confirmed, it is given a number, and later may also be given a name (e.g. 433 Eros). The formal naming convention uses parentheses around the number, but dropping the parentheses is quite common. Informally, it is common to drop the number altogether, or to drop it after the first mention when a name is repeated in running text.

Minor planets that have been given a number but not a name keep their provisional designation, e.g. (29075) 1950 DA. As modern discovery techniques are finding vast numbers of new asteroids, they are increasingly being left unnamed. The earliest discovered to be left unnamed was for a long time (3360) 1981 VA, now 3360 Syrinx; as of September 2008, this distinction is held by (3708) 1974 $FV_1$. On rare occasions, a small object's provisional designation may become used as a name in itself: the still unnamed (15760) 1992 $QB_1$ gave its "name" to a group of Kuiper belt objects which became known as cubewanos.

A few objects are cross-listed as both comets and asteroids, such as 4015 Wilson–Harrington, which is also listed as 107P/Wilson–Harrington.

## Numbering

Minor planets are awarded an official number once their orbits are confirmed. With the increasing rapidity of discovery, these are now six-figure numbers. The switch from five figures to six figures arrived with the publication of the Minor Planet Circular (MPC) of October 19, 2005, which saw the highest numbered minor planet jump from 99947 to 118161.

## Sources for names

The first few asteroids were named after figures from Greek and Roman mythology, but as such names started to dwindle the names of famous people, literary characters, discoverer's wives, children, and even television characters were used.

The first asteroid to be given a non-mythological name was 20 Massalia, named after the Greek name for the city of Marseilles. The first to be given an entirely non-Classical name was 45 Eugenia, named after Empress Eugénie de Montijo, the wife of Napoleon III. For some time only female (or feminized) names were used; Alexander von Humboldt was the first man to have an asteroid named after him, but his name was feminized to 54 Alexandra. This unspoken tradition lasted until 334 Chicago was named; even then, oddly feminised names show up in the list for years after.

As the number of asteroids began to run into the hundreds, and eventually the thousands, discoverers began to give them increasingly frivolous names. The first hints of this were 482 Petrina and 483 Seppina, named after the discoverer's pet dogs. However, there was little controversy about this until 1971, upon the naming of 2309 Mr. Spock (the name of the discoverer's cat). Although the IAU subsequently banned pet names as sources, eccentric asteroid names are still being proposed and accepted, such as 4321 Zero, 6042 Cheshirecat, 9007 James Bond, 13579 Allodd and 24680 Alleven, and 26858 Misterrogers.

A well-established rule is that, unlike comets, minor planets may not be named after their discoverer(s). One way to circumvent this rule has been for astronomers to exchange the courtesy of naming their discoveries after each other. An exception to this rule is 96747 Crespodasilva, which was named after its discoverer, Lucy d'Escoffier Crespo da Silva, because she died shortly after the discovery, at age 22.[16] [17]

Names were adapted to various languages from the beginning. 1 Ceres, *Ceres* being its Anglo-Latin name, was actually named *Cerere*, the Italian form of the name. German, French, Arabic and Hindi use forms similar to the English, whereas Russian uses a form, *Tserera*, similar to the Italian. In Greek the name was translated to Demeter, the Greek equivalent of the Roman goddess Ceres. In the early years, before it started causing conflicts, asteroids named after Roman figures were generally translated in Greek; other examples are Hera for 3 Juno, Hestia for 4 Vesta, Chloris for 8 Flora, and Pistê for 37 Fides. In Chinese, the names are not given the Chinese forms of the deities they are named after, but rather typically have a syllable or two for the character of the deity or person, followed by 'god(dess)' or 'woman' if just one syllable, plus 'star/planet', so that most asteroid names are written with three Chinese characters. Thus Ceres is 'grain goddess planet',[18] Pallas is 'wisdom goddess planet', etc.

### Special naming rules

Minor-planet naming is not always a free-for-all: there are some populations for which rules have developed about the sources of names. For instance, centaurs (orbiting between Saturn and Neptune) are all named after mythological centaurs; Jupiter trojans after heroes from the Trojan War; resonant trans-Neptunian objects after underworld spirits; and non-resonant TNOs after creation deities.

## Physical properties of comets and minor planets

Commission 15[19] of the International Astronomical Union is dedicated to the Physical Study of Comets & Minor Planets.

Archival data on the physical properties of comets and minor planets are found in the PDS Asteroid/Dust Archive.[20] This includes standard asteroid physical characteristics such as the properties of binary systems, occultation timings and diameters, masses, densities, rotation periods, surface temperatures, albedoes, spin vectors, taxonomy, and absolute magnitudes and slopes. In addition, European Asteroid Research Node (E.A.R.N.), an association of asteroid research groups, maintains a Data Base of Physical and Dynamical Properties of Near Earth Asteroids.[21]

Most detailed information is available from Category: Asteroids visited by spacecraft and Category: Comets visited by spacecraft.

## See also

- List of minor planets
- Solar System
- Groups of minor planets
- Small Solar System body
- Plutoid
- Quasi-satellite

## External links

- Minor Planet Center [22]

## References

[1] "Unusual Minor Planets" (http://www.minorplanetcenter.net/iau/lists/Unusual.html). Minor Planet Center. . Retrieved 23 December 2011.

[2] "Minor Planet Statistics" (http://www.minorplanetcenter.net/iau/lists/ArchiveStatistics.html). Minor Planet Center. . Retrieved 2011-02-26.

[3] When did the asteroids become minor planets? (http://aa.usno.navy.mil/faq/docs/minorplanets.php), James L. Hilton, Astronomical Information Center, United States Naval Observatory. Accessed May 5, 2008.

[4] Planet, asteroid, minor planet: A case study in astronomical nomenclature, David W. Hughes, Brian G. Marsden, *Journal of Astronomical History and Heritage* **10**, #1 (2007), pp. 21–30. Bibcode: 2007JAHH...10...21H

[5] " Asteroid (http://encarta.msn.com/encyclopedia_761551567/asteroid.html)", *MSN Encarta*, Microsoft. Accessed May 5, 2008. Archived (http://www.webcitation.org/5kx3nDxZi) 2009-11-01.

[6] Press release, IAU 2006 General Assembly: Result of the IAU Resolution votes (http://www.iau.org/public_press/news/detail/iau0603/), International Astronomical Union, August 24, 2006. Accessed May 5, 2008.

[7] Questions and Answers on Planets (http://www.iau.org/public_press/news/release/iau0603/questions_answers/), additional information, news release IAU0603, IAU 2006 General Assembly: Result of the IAU Resolution votes, International Astronomical Union, August 24, 2006. Accessed May 8, 2008.

[8] JPL. "How Many Solar System Bodies" (http://ssd.jpl.nasa.gov/?body_count). *JPL Solar System Dynamics*. NASA. . Retrieved 2010-09-27.

[9] "Discovery Circumstances: Numbered Minor Planets (1)-(5000)" (http://www.minorplanetcenter.net/iau/lists/NumberedMPs000001.html). Minor Planet Center. . Retrieved 2011-02-26.

[10] "Discovery Circumstances: Numbered Minor Planets (240001)-(245000)" (http://www.minorplanetcenter.net/iau/lists/NumberedMPs240001.html). Minor Planet Center. . Retrieved 2011-02-26.

[11] Yeomans, Don, "Near-Earth Object groups" (http://neo.jpl.nasa.gov/neo/groups.html), *Near Earth Object Project* (NASA), , retrieved 2011-12-24

[12] Connors, Martin; Wiegert, Paul; Veillet, Christian (July 2011), "Earth's Trojan asteroid", *Nature* **475** (7357): 481–483, Bibcode 2011Natur.475..481C, doi:10.1038/nature10233, PMID 21796207

[13] Trilling, David et al. (October 2007), "DDT observations of five Mars Trojan asteroids", *Spitzer Proposal ID #465*, Bibcode 2007sptz.prop..465T

[14] Horner, J.; Evans, N.W.; Bailey, M. E. (2004). "Simulations of the Population of Centaurs I: The Bulk Statistics". *Monthly Notices of the Royal Astronomical Society* **354** (3): 798–810. arXiv:astro-ph/0407400. Bibcode 2004MNRAS.354..798H. doi:10.1111/j.1365-2966.2004.08240.x.

[15] Neptune trojans, Jupiter trojans (http://www.dtm.ciw.edu/users/sheppard/trojans/)

[16] NASA JPL Small-Body Database Browser on 96747 Crespodasilva (http://ssd.jpl.nasa.gov/sbdb.cgi?sstr=96747+Crespodasilva)

[17] Staff (November 28, 2000). "Lucy Crespo da Silva, 22, a senior, dies in fall" (http://hubblesite.org/newscenter/newsdesk/archive/releases/1991/12/text/). Hubble News Desk. . Retrieved 2008-04-15.

[18] 'valley' being a common abbreviation of 'grain' that would be formally adopted with simplified Chinese characters.

[19] "Division III Commission 15 Physical Study of Comets & Minor Planets" (http://www.iau.org/science/scientific_bodies/commissions/15). International Astronomical Union (IAU). September 29, 2005. . Retrieved 2010-03-22.

[20] "Physical Properties of Asteroids" (http://www.psi.edu/pds/archive/physical.html). .

[21] "The Near-Earth Asteroids Data Base" (http://earn.dlr.de/nea). .

[22] http://www.minorplanetcenter.net/iau/mpc.html

# Dwarf_planet

A **dwarf planet** is a planetary-mass object that is neither a planet nor a satellite. More explicitly, the International Astronomical Union (IAU) defines a *dwarf planet* as a celestial body in direct orbit of the Sun[1] that is massive enough for its shape to be controlled by gravitational rather than mechanical forces (that is, it has sufficient mass to overcome its internal compressive strength and achieve hydrostatic equilibrium, and is thus an ellipsoid in shape), but that unlike a planet has not cleared its orbital region of other objects.[2] [3] The term *dwarf planet* was adopted in 2006 as part of a three-way categorization of bodies orbiting the Sun,[1] brought about by an increase in discoveries of trans-Neptunian objects that rivaled Pluto in size, and finally precipitated by the discovery of an even more massive object, Eris.[4] This classification states that bodies large enough to have cleared the neighbourhood of their orbit are defined as *planets*, while those that are not massive enough to be rounded by their own gravity are defined as *small Solar System bodies.* Dwarf planets come in between. The exclusion of dwarf planets from the roster of planets by the IAU has been both praised and criticized; it was said to be the "right decision" by Mike Brown,[5] [6] [7] who discovered Eris and other new dwarf planets, but has been rejected by Alan Stern,[8] [9] who had coined the term *dwarf planet* in 1990.[10]

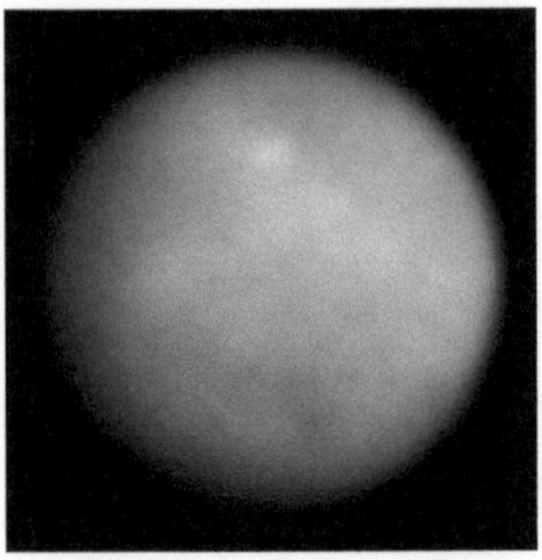

Ceres as seen with the Hubble Space Telescope. It is the only dwarf planet in the asteroid belt.

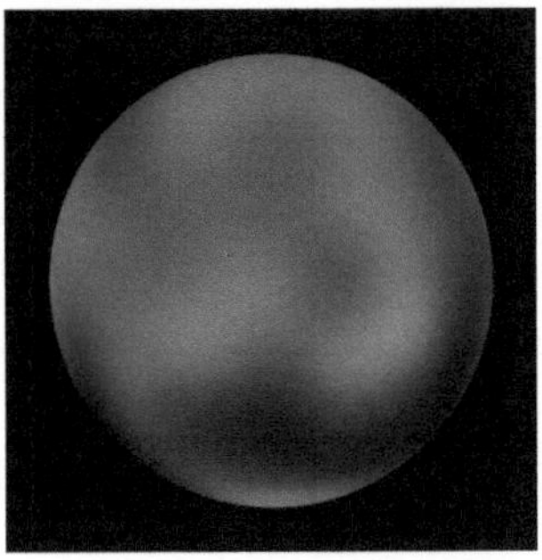

Pluto in approximate true colour based on Hubble Space Telescope albedo data

The IAU currently recognizes five dwarf planets in the Solar System: Ceres, Pluto, Haumea, Makemake, and Eris.[11] However, only two of these bodies, Ceres and Pluto, have been observed in enough detail to demonstrate that they fit the definition. Eris has been accepted as a dwarf planet because it is more massive than Pluto. The IAU subsequently decided that unnamed trans-Neptunian objects with an absolute magnitude brighter than +1 (and hence a diameter of ≥838 km assuming a geometric albedo of ≤1)[12] are to be named under the assumption that they are dwarf planets.[13] The only two such objects known at the time, Makemake and Haumea, went through this naming procedure and were declared to be dwarf planets.

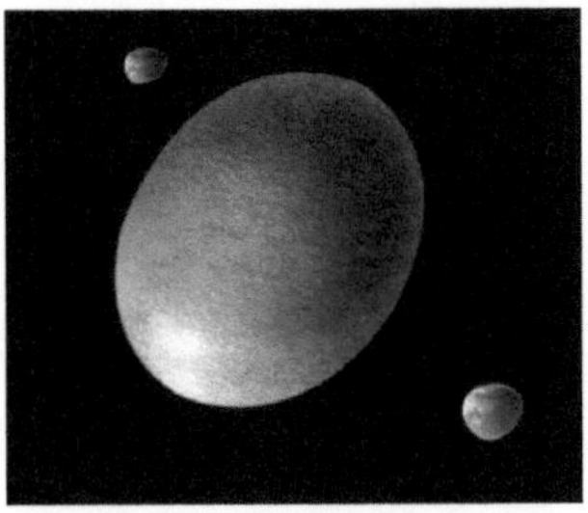

Haumea with its moons, Hi'iaka and Namaka (artist's conception)

It is suspected that another hundred or so known objects in the Solar System are dwarf planets.[14] Estimates are that up to 200 dwarf planets may be found when the entire region known as the Kuiper belt is explored, and that the number may exceed 10,000 when objects scattered outside the Kuiper belt are considered.[15] Individual astronomers have accepted several additional bodies as dwarf planets,[14] [15] and Mike Brown published in August 2011 a list of 390 candidate objects, organized in categories from "nearly certainly" to "possibly" meeting the IAU's criteria, along with his classification methodology.[16] Brown identifies nine known objects – the five accepted by the IAU plus 2007 $OR_{10}$, Sedna, Quaoar, and Orcus – as "virtually certain", with another two dozen highly likely, and there are probably a hundred or so such objects in total.[14]

The classification of bodies in other planetary systems with the characteristics of dwarf planets has not been addressed.[17]

Makemake (artist's conception)

## History of the concept

Before the discoveries of the early 21st century, astronomers had no strong need for a formal definition of a planet. With the discovery of Pluto in 1930, most astronomers considered the Solar System to have nine planets, along with thousands of significantly smaller bodies such as asteroids and comets. For almost 50 years Pluto was thought to be larger than Mercury,[18] [19] but with the discovery in 1978 of Pluto's moon Charon, it became possible to measure Pluto's mass accurately and to determine that it was much smaller than in initial estimates.[20] It was roughly one-twentieth the mass of Mercury, which made Pluto by far the smallest planet. Although it was still more than ten times as massive as the largest object in the asteroid belt, Ceres, it was one-fifth that of Earth's Moon.[21] Furthermore, having some unusual characteristics such as large orbital eccentricity and a high orbital inclination, it became evident it was a completely different kind of body from any of the other planets.[22]

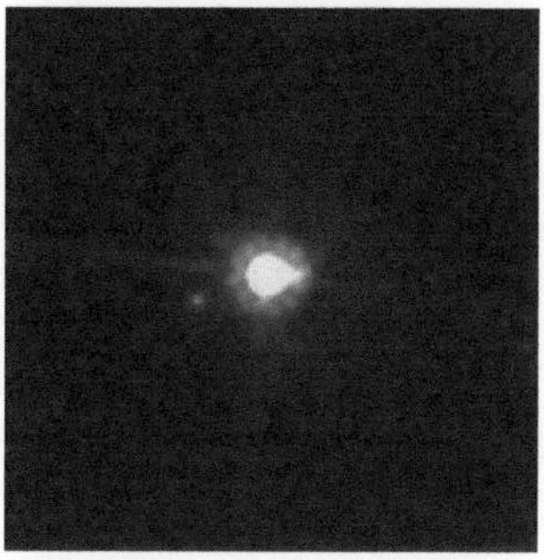
Eris as seen with the Hubble Space Telescope

In the 1990s, astronomers began to find objects in the same region of space as Pluto (now known as the Kuiper belt), and some even farther away.[23] Many of these shared some of the key orbital characteristics of Pluto, and Pluto started being seen as the largest member of a new class of objects, plutinos. This led some astronomers to stop referring to Pluto as a planet. Several terms including *minor planet, subplanet,* and *planetoid* started to be used for the bodies now known as *dwarf planets.*[24] [25] By 2005, three other bodies comparable to Pluto in terms of size and orbit (Quaoar, Sedna, and Eris) had been reported in the scientific literature.[26] It became clear that either they would also have to be classified as planets, or Pluto would have to be reclassified.[27] Astronomers were also confident that more objects as large as Pluto would be discovered, and the number of planets would start growing quickly if Pluto were to remain a planet.[28]

In 2006, Eris (then known as 2003 $UB_{313}$) was believed to be slightly larger than Pluto, and some reports unofficially referred to it as the *tenth planet.*[29] As a consequence, the issue became a matter of intense debate during the IAU General Assembly in August 2006.[30] The IAU's initial draft proposal included Charon, Eris, and Ceres in the list of planets. After many astronomers objected to this proposal, an alternative was drawn up by Uruguayan astronomer Julio Ángel Fernández, in which he created a median classification for objects large enough to be round but that had not cleared their orbits of planetesimals. Dropping Charon from the list, the new proposal also removed Pluto, Ceres, and Eris, since they have not cleared their orbits.[31]

The IAU's final resolution preserved this three-category system for the celestial bodies orbiting the Sun. Fernández suggested calling these median objects *planetoids,*[32] [33] but the IAU's division III plenary session voted unanimously to call them *dwarf planets.*[1] The resolution, #5A, reads:

> The IAU ... resolves that planets and other bodies, except satellites, in our Solar System be defined into three distinct categories in the following way:

(1) A planet[1] is a celestial body that (a) is in orbit around the Sun, (b) has sufficient mass for its self-gravity to overcome rigid body forces so that it assumes a hydrostatic equilibrium (nearly round) shape, and (c) has cleared the neighbourhood around its orbit.

(2) A "*dwarf planet*" is a celestial body that (a) is in orbit around the Sun, (b) has sufficient mass for its self-gravity to overcome rigid body forces so that it assumes a hydrostatic equilibrium (nearly round) shape,[2] (c) has not cleared the neighbourhood around its orbit, and (d) is not a satellite.

(3) All other objects,[3] except satellites, orbiting the Sun shall be referred to collectively as "Small Solar System Bodies."

Footnotes:

[1] The eight planets are: Mercury, Venus, Earth, Mars, Jupiter, Saturn, Uranus, and Neptune.

[2] An IAU process will be established to assign borderline objects either dwarf planet or other status.

[3] These currently include most of the Solar System asteroids, most Trans-Neptunian Objects (TNOs), comets, and other small bodies.

The term *dwarf planet* has itself been somewhat controversial, as a grammatical reading suggests these bodies are planets. The IAU's Resolution 5A was accompanied by a second resolution, 5B, which defined dwarf planets as a subtype of planet, distinguished from the other eight which were to be called "classical planets". Under this arrangement, the twelve planets of the rejected proposal were to be preserved in a distinction between eight "classical planets" and four "dwarf planets". However, resolution 5B was defeated in the same session that 5A was passed, so that only the *dwarf planet* half of the proposal was made official.[34] Because of the grammatical inconsistency of a *dwarf planet* not being a planet, while a *dwarf star* is a star, alternative proposals such as *nanoplanet* and *subplanet* were suggested. However, it was judged that the term *dwarf planet* had already entered general use and that it was too late to change it.[35] The term is parallel to *minor planet*, which is also not a type of planet.

In most languages equivalent terms have been created by translating *dwarf planet* more-or-less literally: French *planète naine*, German *Zwergplanet*, Russian карликовая планета *karlikovaya planeta*, Arabic كوكب قزم *kaukab qazm*, Chinese *ǎixíngxīng*, etc., but Japanese is an exception: In Japanese these bodies are called *junwakusei*
, where *wakusei* is 'planet' and *jun-* is a prefix corresponding to English *quasi-*, *pene-* (almost), and
*sub-*. Thus in Japanese they are called 'subplanets' or 'almost-planets'. The modern Latin name for this class of objects, *planetulus*, or *planetion* following a Greek-styled name, is also a diminutive form of *planeta*, hence meaning "small planet".

Although concerns were raised about the classification of planets orbiting other stars,[17] the issue was not resolved; it was proposed instead to decide this only when such objects start being observed.[31]

The 2006 IAU's Resolution 6a[36] recognizes Pluto as "the prototype of a new category of trans-Neptunian objects". The name and precise nature of this category were not specified but left for the IAU to establish at a later date; in the debate leading up to the resolution, the members of the category were variously referred to as *plutons* and *plutonian objects* but neither name was carried forward.[1] On June 11, 2008, the IAU Executive Committee announced a name, *plutoid*, and a definition: all trans-Neptunian dwarf planets are plutoids,[13] though "in part because of an email miscommunication, the WG-PSN [Working Group for Planetary System Nomenclature] was not involved in choosing the word plutoid. ... In fact, a vote taken by the WG-PSN subsequent to the Executive Committee meeting has rejected the use of that specific term."[37] On July 18, 2008, the WG-PSN classified the object then known as *(136472) 2005 FY*$_9$ as a dwarf planet, and named it Makemake; this was followed in September by Haumea.[38]

# Characteristics

### Planetary discriminants[39]

| Body | Mass ($M_E$*) | $\Lambda$** | $\mu$*** |
|---|---|---|---|
| Mercury | 0.055 | $1.95 \times 10^3$ | $9.1 \times 10^4$ |
| Venus | 0.815 | $1.66 \times 10^5$ | $1.35 \times 10^6$ |
| Earth | 1 | $1.53 \times 10^5$ | $1.7 \times 10^6$ |
| Mars | 0.107 | $9.42 \times 10^2$ | $1.8 \times 10^5$ |
| Ceres | 0.000 15 | $8.32 \times 10^{-4}$ | 0.33 |
| Jupiter | 317.7 | $1.30 \times 10^9$ | $6.25 \times 10^5$ |
| Saturn | 95.2 | $4.68 \times 10^7$ | $1.9 \times 10^5$ |
| Uranus | 14.5 | $3.85 \times 10^5$ | $2.9 \times 10^4$ |
| Neptune | 17.1 | $2.73 \times 10^5$ | $2.4 \times 10^4$ |
| Pluto | 0.002 2 | $2.95 \times 10^{-3}$ | 0.077 |
| Haumea | 0.000 67 | $2.68 \times 10^{-4}$ | 0.02 |
| Makemake | 0.000 67 | $2.22 \times 10^{-4}$ | 0.02[40] |
| Eris | 0.002 8 | $2.13 \times 10^{-3}$ | 0.10 |

*$M_E$ in Earth masses.

**$\Lambda = k\,M^2\,a^{-3/2}$,

where $k = 0.0043$ for units of Yg and AU. $\Lambda > 1$ for planets.[41]

***$\mu = M/m$, where $M$ is the mass of the body,

and $m$ is the aggregate mass of all the other bodies

that share its orbital zone. $\mu > 100$ for planets.

## Orbital dominance

Alan Stern and Harold F. Levison introduced a parameter Λ (lambda), expressing the likelihood of an encounter resulting in a given deflection of orbit.[41] The value of this parameter in Stern's model is proportional to the square of the mass and inversely proportional to the period. Following the authors, this value can be used to estimate the capacity of a body to clear the neighbourhood of its orbit, where $\Lambda > 1$ will eventually clear it. A gap of five orders of magnitude in Λ was found between the smallest terrestrial planets and the largest asteroids and Kuiper belt objects.[39]

Using this parameter, Steven Soter and other astronomers argued for a distinction between planets and dwarf planets based on the inability of the latter to "clear the neighbourhood around their orbits": planets are able to remove smaller bodies near their orbits by collision, capture, or gravitational disturbance (or establish orbital resonances that prevent collisions), while dwarf planets lack the mass to do so.[41] Soter went on to propose a parameter he called the *planetary discriminant*, designated with the symbol μ (mu), that represents an experimental measure of the actual degree of cleanliness of the orbital zone (where μ is calculated by dividing the mass of the candidate body by the total mass of the other objects that share its orbital zone), where $\mu > 100$ is deemed to be cleared.[39] There are several other schemes that try to differentiate between planets and dwarf planets,[8] but the 2006 definition uses this concept.[1]

## Size and mass

Sufficient internal pressure, caused by the body's gravitation, will turn a body plastic, and sufficient plasticity will allow high elevations to sink and hollows to fill in, a process known as gravitational relaxation. Bodies smaller than a few kilometers are dominated by non-gravitational forces and tend to be angular in shape. Larger objects, where gravitation is significant but not dominant, are "potato" shaped; the more massive the body is, the higher its internal pressure and the more rounded its shape, until it achieves maximum rounding at hydrostatic equilibrium. This is the defining limit of a dwarf planet.[42]

When an object is in hydrostatic equilibrium, a global layer of liquid covering its surface would form a liquid surface of the same shape as the body, apart from small-scale surface features such as craters and fissures. If the body does not rotate, it will be a sphere, but the faster it does rotate, the more oblate or even scalene it becomes. However, if such a rotating body were to be heated until it melted, its overall shape would not change when liquid. The extreme example of a non-spherical body in hydrostatic equilibrium is Haumea, which is twice as long along its major axis as it is at the poles. If the body has a massive nearby companion, then tidal forces come into effect as well, distorting it into a prolate spheroid. An example of this is Jupiter's moon Io, which is the most volcanically active body in the Solar System due to effects of tidal heating. Tidal forces also cause a body's rotation to gradually become tidally locked, such that it always presents the same face to its companion. An extreme example of this is the Pluto–Charon system, where both bodies are tidally locked to each other. Earth's Moon is also tidally locked, as are many satellites of the gas giants.

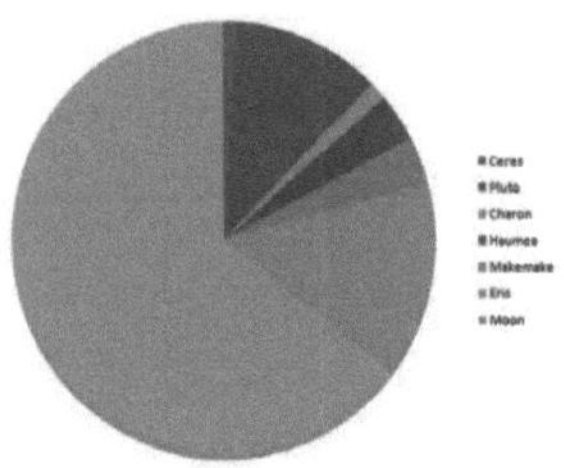

The masses of the four largest plutoids, plus Ceres and Charon, relative to the Earth's Moon. The mass of Makemake is a rough estimate. (See plutoid for a graph of several additional likely dwarf planets without Ceres.)

The upper and lower size and mass limits of dwarf planets have not been specified by the IAU. There is no defined upper limit, and an object larger or more massive than Mercury that has not "cleared the neighbourhood around its orbit" would be classified as a dwarf planet.[43] The lower limit is determined by the requirements of achieving a hydrostatic equilibrium shape, but the size or mass at which an object attains this shape depends on its composition and thermal history. The original draft of the 2006 IAU resolution redefined hydrostatic equilibrium shape as applying "to objects with mass above $5\times10^{20}$ kg and diameter greater than 800 km",[17] but this was not retained in the final draft.[1]

Empirical observations suggest that the lower limit will vary according to the composition and thermal history of the object. For a body made of rigid silicates, such as the stony asteroids, the transition to hydrostatic equilibrium should occur at a diameter of approximately 600 km and a mass of some $3.4\times10^{20}$ kg. For a body made of less rigid water ice, the limit should be about 320 km and $10^{19}$ kg. [44] In the asteroid belt, Ceres is the only body that clearly surpasses the silicaceous limit (though it is actually a rocky–icy body), and its shape is an equilibrium spheroid. 2 Pallas and 4 Vesta, however, are rocky and are just below the limit. Pallas, at 525–560 km and $1.85–2.4\times10^{20}$ kg, is "nearly round" but still somewhat irregular. Vesta, at 530 km and $2.6\times10^{20}$ kg, deviates from an ellipsoid shape primarily due to a large impact basin at its pole.

Among icy bodies, the smallest known to be in hydrostatic equilibrium is Mimas, at 396 km and $3.75\times10^{19}$ kg. The largest irregular body in the outer Solar System is Proteus, nearly-but-not-quite round at 405–435 km and an assumed mass of $\approx4.4\times10^{19}$ kg. Bodies like Mimas may have had a warmer thermal history than Proteus, or their shape may have resolved after a collision.[45] Neither body is pure ice as used to calculate the lowest limit, however, and Mike Brown suggests that the practical lower limit for an icy dwarf planet is likely to be somewhere under 400 km.[46] There are about 100 TNOs currently estimated to be above this size.

## Official and nearly certain dwarf planets

As of 2011, five objects are recognized as dwarf planets by the IAU.[47] Mike Brown considers these and four others to be "nearly certain".[14] For Ceres and Pluto, this is known through direct observation.[48] Brown thinks the other seven are massive enough to be in hydrostatic equilibrium even if they are dense (primarily rocky) and at the lower end of their estimated diameters.[14] Eris is more massive than Pluto; Haumea and Makemake were assigned names as dwarf planets based on their absolute magnitudes.[11] [36] In relative distance from the Sun, the five "official" dwarf planets are:

1. Ceres ⚳ – discovered on January 1, 1801, 45 years before Neptune. Considered a planet for half a century before reclassification as an asteroid. Accepted as a dwarf planet by the IAU on September 13, 2006.
2. Pluto ♇ – discovered on February 18, 1930. Classified as a planet for 76 years. Reclassified as a dwarf planet by the IAU on August 24, 2006.
3. Haumea – discovered on December 28, 2004. Accepted by the IAU as a dwarf planet on September 17, 2008.
4. Makemake – discovered on March 31, 2005. Accepted by the IAU as a dwarf planet on July 11, 2008.
5. Eris – discovered on January 5, 2005. Called the "tenth planet" in media reports. Accepted by the IAU as a dwarf planet on September 13, 2006.

The other four "nearly certain" objects per Brown, Tancredi, et al. are:

1. Orcus – discovered on February 17, 2004.
2. Quaoar – discovered on June 5, 2002.
3. 2007 $OR_{10}$ – discovered on July 17, 2007.
4. Sedna – discovered on November 14, 2003.

No space probes have visited any of these. This will change if NASA's *Dawn* and *New Horizons* missions reach Ceres and Pluto, respectively. The plans are for *Dawn* to orbit Ceres, and *New Horizons* to fly by Pluto, both in 2015. *Dawn* entered orbit around the *potential* dwarf planet Vesta on 16 July 2011.[49] On 5 September 2012, *Dawn* left Vesta in order to orbit Ceres, which it is scheduled to reach in February 2015.[50] [51]

### Objects recognized by the IAU as dwarf planets

| Orbital attributes[52] | | | | | | | |
|---|---|---|---|---|---|---|---|
| **Name** | **Region of Solar System** | **Orbital radius (AU)** | **Orbital period (years)** | **Mean orbital speed (km/s)** | **Inclination to ecliptic** | **Orbital eccentricity** | **Planetary discriminant** |
| **Ceres** | Asteroid belt | 2.77 | 4.60 | 17.882 | 10.59° | 0.079 | 0.33 |
| **Pluto** | Kuiper belt (plutino) | 39.48 | 248.09 | 4.666 | 17.14° | 0.249 | 0.077 |
| **Haumea** | Kuiper belt (12:7) | 43.13 | 283.28 | | 28.22° | 0.195 | 0.020 |
| **Makemake** | Kuiper belt (cubewano) | 45.79 | 309.9 | 4.419 | 28.96° | 0.159 | 0.02 |
| **Eris** | Scattered disc | 67.67 | 557 | 3.436 | 44.19° | 0.442 | 0.10 |

| Physical attributes | | | | | | | | | | | | |
|---|---|---|---|---|---|---|---|---|---|---|---|---|
| Name | Equatorial diameter relative to the Moon | Equatorial diameter (km)[14] | Mass relative to the Moon | Mass ($\times 10^{21}$ kg) | Density (g/cm$^3$) | Surface gravity (m/s$^2$) | Escape velocity (km/s) | Axial inclination | Rotation period (days) | Moons | Surface temp. (K) | Atmosphere |
| **Ceres** | 28% | 974.6±3.2 | 1.3% | 0.94 | 2.08 | 0.27 | 0.51 | ≈ 3° | 0.38 | 0 | 167 | none |
| **Pluto** | 66% | 2306±20 | 17.8% | 13.05 | 2.0 | 0.58 | 1.2 | 119.59° | −6.39 | 5 | 44 | transient |
| **Haumea** | ≈ 37% | 1300±? | 5.5% | 4.01 ± 0.04 | 2.6–3.3 (?) | 0.44 | 0.84 | | 0.16 | 2 | 32 ± 3 | ? |
| **Makemake** | 41% | 1420±60 | ≈ 4% ? | ≈ 3 ? | 1.7 ± 0.3[53] | | | ? | 0.32 | 0 | ≈ 30 | none[53] |
| **Eris** | 67% | 2326±12 | 22.7% | 16.7 | 2.5 | ≈ 0.8 | 1.3 | | ≈ 1 (0.75–1.4) | 1 | ≈ 42 | transient? |

# Objects recognized by others as dwarf planets

| Orbital attributes[52] | | | | | | | |
|---|---|---|---|---|---|---|---|
| Name | Region of Solar System | Orbital radius (AU) | Orbital period (years) | Mean orbital speed (km/s) | Inclination to ecliptic | Orbital eccentricity | Planetary discriminant |
| **Orcus** | Kuiper belt (plutino) | 39.17 | 245.18 | | 20.57° | 0.227 | 0.003 |
| **Quaoar** | Kuiper belt (cubewano) | 43.405 | 285.97 | | 8.00° | 0.039 | 0.007–0.010 |
| **2007 OR$_{10}$** | Scattered disc (10:3?) | 67.21 | 550.98 | | 30.70° | 0.500 | ? |
| **Sedna** | Detached | 518.57 | ≈11,400 | | 11.93° | 0.853 | ? |

| Physical attributes | | | | | | | | | | | | |
|---|---|---|---|---|---|---|---|---|---|---|---|---|
| Name | Equatorial diameter relative to the Moon | Equatorial diameter (km) | Mass relative to the Moon | Mass ($\times 10^{21}$ kg) | Density (g/cm$^3$) | Surface gravity (m/s$^2$) | Escape velocity (km/s) | Axial inclination | Rotation period (days) | Moons | Surface temp. (K) | Atmosphere |
| **Orcus** | ≈ 22% | 760–810 | 0.9% | 0.63 | | | | ° | 0.55 | 1 | | |
| **Quaoar** | ≈ 26% | 890±70 | 1.8–2.6% | 1.6 ± 0.3 | | | | ° | 0.74 | 1 | | |
| **2007 OR$_{10}$** | ≈ 37% | 1280±210 | | ? | | | | ? | ? | 0 | | |
| **Sedna** | ≈ 30% | 995 ± 80 | ≈1.4% | ≈1 | | | | ? | 0.42 | 0 | ≈ 12 | |

## Additional possibilities

After Ceres, the next-most-massive body in the asteroid belt, Vesta, might also be classified as a dwarf planet, as its shape appears to deviate from hydrostatic equilibrium mainly because of massive impacts that occurred after it solidified.[54] The definition of dwarf planet does not address this issue. Data from the *Dawn* probe, which orbited Vesta in 2011–2012, may help clarify matters.[55]

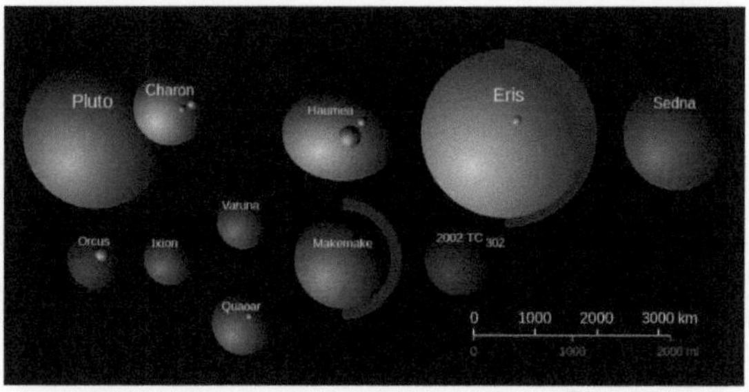

Illustration of the relative sizes, albedos, and colours of the largest trans-Neptunian objects

Many trans-Neptunian objects (TNOs) are thought to have icy cores and therefore would require a diameter of perhaps 400 km (250 mi)—only about 3% of that of Earth—to relax into gravitational equilibrium, making them dwarf planets of the plutoid class.[46] Although only rough estimates of the diameters of these objects are available, as of 2011 it was believed that a hundred of the known bodies beyond Neptune were probable dwarf planets.[14] A team is investigating thirty of these, and believe that the number will eventually prove to be about 200 in the Kuiper belt and many more beyond.[46]

## Planetary-mass moons

Nineteen moons are known to be massive enough to have relaxed into a rounded shape under their own gravity, and seven of them are more massive than either Eris or Pluto. They are not physically distinct from the dwarf planets, but are not members of that class because they do not directly orbit the Sun. The seven which are more massive than Eris are Earth's moon, the four Galilean moons of Jupiter (Io, Europa, Ganymede, and Callisto), Titan of Saturn, and one moon of Neptune (Triton). The others are six moons of Saturn (Mimas, Enceladus, Tethys, Dione, Rhea, and Iapetus), five moons of Uranus (Miranda, Ariel, Umbriel, Titania, and Oberon), and one moon of Pluto (Charon). The term *planemo* ("planetary-mass object") covers both dwarf planets and such moons, as well as planets.[56] Alan Stern considers them a special category of planets, "satellite planets".[57]

Despite its requirement that dwarf planets orbit the Sun directly, draft resolution (5) presented to the IAU, which considered dwarf planets to be a category of planet, stated Charon could be considered a planet because it revolves with Pluto around a common center of mass located between the two bodies (rather than within one of the bodies).[58] [17] Note, however, that the Jupiter–Sun centre of mass lies outside the Sun, and so such a definition would imply that Jupiter is not a satellite of the Sun and therefore not a planet. This definition was not preserved in the IAU's final resolution.

## Contention

In the immediate aftermath of the IAU definition of dwarf planet, a number of scientists expressed their disagreement with the IAU resolution.[8] Campaigns included car bumper stickers and T-shirts.[59] Mike Brown (the discoverer of Eris) agrees with the reduction of the number of planets to eight.[60]

NASA has announced that it will use the new guidelines established by the IAU.[61] However, Alan Stern, the director of NASA's mission to Pluto, rejects the current IAU definition of planet, both in terms of defining dwarf planets as something other than a type of planet, and in using orbital characteristics (rather than intrinsic characteristics) of objects to define them as dwarf planets.[62] Thus, as of 2011, he still referred to Pluto as a planet,[63] and accept other dwarf planets such as Ceres and Eris, as well as the larger moons, as additional planets.[57] Several years before the IAU definition, he used orbital characteristics to separate "überplanets" (the dominant eight) from "unterplanets" (the dwarf planets), considering both 'planets'.[41]

## See also

- List of planetary bodies
- Lists of small Solar System bodies
- List of Solar System bodies formerly regarded as planets
- List of Solar System objects in hydrostatic equilibrium
- Mesoplanet

## Notes

[1] "Definition of a Planet in the Solar System: Resolutions 5 and 6" (http://www.iau.org/static/resolutions/Resolution_GA26-5-6.pdf) (PDF). *IAU 2006 General Assembly* (International Astronomical Union). 2006-08-24. . Retrieved 2008-01-26.

[2] "IAU 2006 General Assembly: Result of the IAU Resolution votes" (http://www.iau.org/iau0603.414.0.html). International Astronomical Union. 2006. . Retrieved 2008-01-26.

[3] "Dwarf Planets" (http://solarsystem.nasa.gov/planets/profile.cfm?Object=Dwarf). NASA. . Retrieved 2008-01-22.

[4] Error: Bad DOI specified!

[5] Koski, Olivia (2010-12-27). "Q&A: Astronomer Mike Brown on How He Killed Pluto" (http://www.wired.com/magazine/2010/12/pl_print_mikebrown/). Wired. . Retrieved 2012-02-12.

[6] Perlman, David (2006-08-25). "Pluto demoted -- from 9th planet to just a dwarf" (http://astro.berkeley.edu/~basri/defineplanet/Chron06.htm). *San Francisco Chronicle*. . Retrieved 2012-02-12.

[7] Kennedy, Stephanie (2006-08-25). "Pluto stripped of planet status" (http://www.abc.net.au/am/content/2006/s1724061.htm). *"AM", ABC Local Radio*. . Retrieved 2012-02-12.

[8] Rincon, Paul (2006-08-25). "Pluto vote 'hijacked' in revolt" (http://news.bbc.co.uk/2/hi/science/nature/5283956.stm). BBC News. . Retrieved 2008-01-26.

[9] Jorge Salazar (2009-11-30). "Alan Stern: 'A Chihuahua is still a dog, and Pluto is still a planet'" (http://www.earthsky.org/interviewpost/space/alan-stern-âa-chihuahua-is-still-a-dog-and-pluto-is-still-a-planet). EarthSky (Earthsky Interviews). . Retrieved 2009-12-08.

[10] S. Alan Stern, "On the number of planets in the outer solar system: Evidence of a substantial population of 1000-km bodies", *Icarus* 90:2, April 1991

[11] "Dwarf Planets and their Systems" (http://planetarynames.wr.usgs.gov/append7.html#DwarfPlanets). Working Group for Planetary System Nomenclature (WGPSN). 2008-07-11. . Retrieved 2008-07-13.

[12] Dan Bruton. "Conversion of Absolute Magnitude to Diameter for Minor Planets" (http://www.physics.sfasu.edu/astro/asteroids/sizemagnitude.html). Department of Physics & Astronomy (Stephen F. Austin State University). . Retrieved 2008-06-13.

[13] "Plutoid chosen as name for Solar System objects like Pluto" (http://www.iau.org/public_press/news/release/iau0804/) (Press release). .

[14] Mike Brown, 'How many dwarf planets are there in the outer solar system?' (http://www.gps.caltech.edu/~mbrown/dps.html) Accessed 2011-08-24

[15] "Today we know of more than a dozen dwarf planets in the solar system". The PI's Perspective (http://pluto.jhuapl.edu/overview/piPerspective.php?page=piPerspective_08_24_2012)

[16] "Free the Dwarf Planets!" (http://www.mikebrownsplanets.com/2011/08/free-dwarf-planets.html). Michael Brown. 2011-08-24. . Retrieved 2011-08-24.

[17] "The IAU draft definition of "planet" and "plutons"" (http://www.iau.org/public_press/news/release/iau0601/). International Astronomical Union. 2006-08-16. . Retrieved 2008-05-17.

[18] Mager, Brad. "Pluto Revealed" (http://www.discoveryofpluto.com/pluto06.html). discoveryofpluto.com. . Retrieved 2008-01-26.

[19] Cuk, Matija; Masters, Karen (2007-09-14). "Is Pluto a planet?" (http://curious.astro.cornell.edu/question.php?number=624). Cornell University, Astronomy Department. . Retrieved 2008-01-26.

[20] Buie, Marc W.; William M. Grundy, Eliot F. Young, Leslie A. Young and S. Alan Stern (2006). "Orbits and Photometry of Pluto's Satellites: Charon, S/2005 P1, and S/2005 P2" (http://www.iop.org/EJ/abstract/1538-3881/132/1/290). *The Astronomical Journal* **132** (132): 290–98. arXiv:astro-ph/0512491. Bibcode 2006AJ....132..290B. doi:10.1086/504422. . Retrieved 2008-02-10.

[21] Jewitt, David; Delsanti, Audrey (2006) (PDF). *The Solar System Beyond The Planets in Solar System Update : Topical and Timely Reviews in Solar System Sciences (PDF)* (http://www.ifa.hawaii.edu/faculty/jewitt/papers/2006/DJ06.pdf). Springer. doi:10.1007/3-540-37683-6. ISBN 978-3-540-37683-5. . Retrieved 2008-02-10.

[22] Weintraub, David A. (2006). *Is Pluto a Planet? A Historical Journey through the Solar System*. Princeton, N.J.: Princeton Univ. Press. pp. 1–272. ISBN 978-0-691-12348-6.

[23] Phillips, Tony; Phillips, Amelia (2006-09-04). "Much Ado about Pluto" (http://www.plutopetition.com/unplanet.php). PlutoPetition.com. . Retrieved 2008-01-26.

[24] "Planetoids Beyond Pluto" (http://www.astrobio.net/news/modules.php?op=modload&name=News&file=article&sid=1366). Astrobiology Magazine. 2004-12-30. . Retrieved 2008-01-26.

[25] "Hubble Observes Planetoid Sedna, Mystery Deepens" (http://hubblesite.org/newscenter/archive/releases/2004/14/). NASA's Hubble Space Telescope home site. 2004-04-14. . Retrieved 2008-01-26.

[26] Brown, Michael E.. "The Discovery of Eris, the Largest Known Dwarf Planet" (http://www.gps.caltech.edu/~mbrown/planetlila/). California Institute of Technology, Department of Geological Sciences. . Retrieved 2008-01-26.

[27] Brown, Michael E. (2004). "What is the definition of a planet?" (http://web.gps.caltech.edu/~mbrown/sedna/index.html#planets). California Institute of Technology, Department of Geological Sciences. . Retrieved 2008-01-26.

[28] Brown, Mike (2006-08-16). "War of the Worlds" (http://www.nytimes.com/2006/08/16/opinion/16brown.html). New York Times. . Retrieved 2008-02-20.

[29] "Astronomers Measure Mass of Largest Dwarf Planet" (http://hubblesite.org/newscenter/archive/releases/2007/24/full/). NASA's Hubble Space Telescope home site. 2007-06-14. . Retrieved 2008-01-26.

[30] Brown, Michael E.. "What makes a planet?" (http://www.gps.caltech.edu/~mbrown/whatsaplanet/). California Institute of Technology, Department of Geological Sciences. . Retrieved 2008-01-26.

[31] Britt, Robert Roy (2006-08-19). "Details Emerge on Plan to Demote Pluto" (http://www.space.com/scienceastronomy/060819_new_proposal.html). Space.com. . Retrieved 2006-08-18.

[32] Bailey, Mark E.. "Comments & discussions on Resolution 5: The definition of a planet – Planets Galore" (http://astro.cas.cz/nuncius/appendix.html). *Dissertatio cum Nuncio Sidereo, Series Tertia – official newspaper of the IAU General Assembly 2006*. Astronomical Institute Prague. . Retrieved 2008-02-09.

[33] "Dos uruguayos, Julio Fernández y Gonzalo Tancredi en la historia de la astronomía:reducen la cantidad de planetas de 9 a 8 ...&Anotaciones de Tancredi" (http://web.archive.org/web/20071220063342/http://www.ici.edu.uy/perfilASTRO.htm) (in Spanish). Science and Research Institute, Mercedes, Uruguay. Archived from the original (http://www.ici.edu.uy/perfilASTRO.htm) on December 20, 2007. . Retrieved 2008-02-11.

[34] Mike Brown, 2010. *How I Killed Pluto and Why It Had It Coming.*

[35] IAU (2009). Reports on Astronomy 2006–2009 (http://journals.cambridge.org/article_S1743921308025398). *Transactions of the IAU*, vol. XXVII-A

[36] "IAU 2006 General Assembly: Result of the IAU Resolution votes" (http://www.iau.org/public_press/news/release/iau0603). .

[37] IAU (2009). Division III (Planetary Systems Sciences): Triennial Report 2006–2009 (http://journals.cambridge.org/article_S1743921308025398). *Transactions IAU, Volume XXVIIA*

[38] "Minor Planets, Dwarf Planets" (http://cbat.eps.harvard.edu/Headlines.html). IAU. . Retrieved 2010-10-24.

[39] Soter, Steven (2006-08-16). "What is a Planet?". *The Astronomical Journal* **132** (6): 2513–19. arXiv:astro-ph/0608359. Bibcode 2006AJ....132.2513S. doi:10.1086/508861.

[40] Calculated using the estimate for the mass of the Kuiper belt found in Iorio, 2007 (http://adsabs.harvard.edu/abs/2007MNRAS.tmp...24I) of 0.033 Earth masses

[41] Stern, S. Alan; Levison; and Levison, Harold F. (2002). "Regarding the criteria for planethood and proposed planetary classification schemes" (http://www.boulder.swri.edu/~hal/PDF/planet_def.pdf) (PDF). *Highlights of Astronomy* **12**: 205–213, as presented at the XXIVth General Assembly of the IAU–2000 [Manchester, UK, 7–18 August 2000]. Bibcode 2002HiA....12..205S. .

[42] Lineweaver & Marc Norman, 2010, "The Potato Radius: a Lower Minimum Size for Dwarf Planets"

[43] Indeed, Mike Brown has set out to find such an object. ( "Julia Sweeney and Michael E. Brown" (http://web.archive.org/web/20080626220548/http://www.pluggd.tv/audio/channels/kcet_podcast__hammer_conversations/episodes/2h10l). *Hammer Conversations: KCET podcast*. 2007. Archived from the original (http://www.pluggd.tv/audio/channels/kcet_podcast__hammer_conversations/episodes/2h10l) on 2008-06-26. . Retrieved 2008-06-28.)

[44] G.H.A. Cole, 2000, Minimum Radius And Mass For A Planetary Body (http://web.archive.org/web/20110606080126/http://www.hull.ac.uk/php/dmsghc/planets/html/v_minimum_radius_and_mass_for_a_planetary_body.htm), archived from http://www.hull.ac.uk/php/dmsghc/planets/html/v_minimum_radius_and_mass_for_a_planetary_body.htm

[45] McKinnon, William B. *et al.* (2008). "Structure and Evolution of Kuiper Belt Objects and Dwarf Planets" (http://www.lpi.usra.edu/books/ssbn2008/7035.pdf). In Barucci, M. A. *et al.*. *The Solar System Beyond Neptune*. Tucson: University of Arizona Press. p. 220. Bibcode 2008ssbn.book..213M. ISBN 978-0-8165-2755-7. .

[46] Brown, Michael E.. "The Dwarf Planets" (http://web.gps.caltech.edu/~mbrown/dwarfplanets/). California Institute of Technology, Department of Geological Sciences. . Retrieved 2008-01-26.

[47] "IAU names fifth dwarf planet Haumea" (http://www.iau.org/public_press/news/detail/iau0807/). Paris: International Astronomical Union. 2008-09-17. . Retrieved 16 September 2011.

[48] Error: Bad DOI specified!

[49] "DAWN: Mission status" (http://dawn.jpl.nasa.gov/mission/status.asp). NASA/JPL. . Retrieved 2011-07-25.

[50] "NASA's Dawn Spacecraft Hits Snag on Trip to 2 Asteroids" (http://www.space.com/17119-nasa-dawn-asteroid-spacecraft-vesta.html). Space.com. August 15, 2012. . Retrieved August 27, 2012.

[51] "Dawn Gets Extra Time to Explore Vesta" (http://www.nasa.gov./mission_pages/dawn/news/dawn20120418.html). NASA. 2012-04-18. . Retrieved 2012-04-24.

[52] Bowell, Ted. "The Asteroid Orbital Elements Database" (ftp://ftp.lowell.edu/pub/elgb/astorb.html). Lowell Observatory. . Retrieved 2008-02-12.

[53] European Southern Observatory. "Dwarf Planet Makemake Lacks Atmosphere: Distant Frigid World Reveals Its Secrets for First Time" (http://www.sciencedaily.com/releases/2012/11/121121145516.htm). ScienceDaily. . Retrieved 19 November 2012.

[54] Thomas, Peter C.; Binzelb, Richard P.; Gaffeyc, Michael J.; Zellnerd, Benjamin H.; Storrse, Alex D.; Wells, Eddie (1997). "Vesta: Spin Pole, Size, and Shape from HST Images". *Icarus* **128** (1): 88–94. Bibcode 1997Icar..128...88T. doi:10.1006/icar.1997.5736.

[55] Russel, C.T.; Capaccioni, F.; Coradini, A.; et al. (2006). "Dawn Discovery mission to Vesta and Ceres: Present status". *Advances in Space Research* **38** (9): 2043–48. Bibcode 2006AdSpR..38.2043R. doi:10.1016/j.asr.2004.12.041.

[56] Basri, G.; Brown, M.E. (2006). "Planetesimals to Brown Dwarfs: What is a Planet?" (http://siba.unipv.it/fisica/articoli/A/Annual Review Earth Planetary Sciences_vol.34_2006_pp.193-216.pdf) (PDF). *Annual Review of Earth and Planetary Sciences* **34**: 193–216. arXiv:astro-ph/0608417. Bibcode 2006AREPS..34..193B. doi:10.1146/annurev.earth.34.031405.125058. .

[57] "Should Large Moons Be Called 'Satellite Planets'?" (http://news.discovery.com/space/should-large-moons-be-called-satellite-planets.html). News.discovery.com. 2010-05-14. . Retrieved 2011-11-04.

[58] The footnote in the original text reads: *For two or more objects comprising a multiple object system.... A secondary object satisfying these conditions i.e. that of mass, shape is also designated a planet if the system barycentre resides outside the primary. Secondary objects not satisfying these criteria are "satellites". Under this definition, Pluto's companion Charon is a planet, making Pluto–Charon a double planet.*

[59] Chang, Alicia (2006-08-25). "Online merchants see green in Pluto news" (http://www.usatoday.com/tech/science/space/2006-08-25-pluto-memorabilia_x.htm). *Associated Press* (USA Today). . Retrieved 2008-01-25.

[60] Brown, Michael E.. "The Eight Planets" (http://web.gps.caltech.edu/~mbrown/eightplanets/). California Institute of Technology, Department of Geological Sciences. . Retrieved 2008-01-26.

[61] "Hotly-Debated Solar System Object Gets a Name" (http://www.nasa.gov/vision/universe/solarsystem/erisf-20060914.html). NASA press release. 2006-09-14. . Retrieved 2008-01-26.

[62] Stern, Alan (2006-09-06). "Unabashedly Onward to the Ninth Planet" (http://pluto.jhuapl.edu/overview/piPerspectives/piPerspective_09_06_2006.php). New Horizons Web Site. . Retrieved 2008-01-26.

[63] Wall, Mike (2011-08-24). "Pluto's Planet Title Defender: Q & A With Planetary Scientist Alan Stern" (http://www.space.com/12710-pluto-defender-alan-stern-dwarf-planet-interview.html). SPACE.com. . Retrieved 2012-12-03.

## References

## External links

- NPR: Dwarf Planets May Finally Get Respect (http://www.npr.org/templates/story/story.php?storyId=5631291) (David Kestenbaum)
- BBC News: Q&A New planets proposal (http://news.bbc.co.uk/1/hi/sci/tech/4798205.stm), August 16, 2006
- Ottawa Citizen: The case against Pluto (http://www.canada.com/ottawacitizen/news/story.html?id=a9591aed-f19f-4ac3-a324-1f8bb46d9379&p=2) (P. Surdas Mohit) August 24, 2006
- James L. Hilton, When Did the asteroids Become Minor Planets? (http://aa.usno.navy.mil/faq/docs/minorplanets.php)
- NASA: IYA 2009 Dwarf Planets (http://astronomy2009.nasa.gov/topics_sep.htm)

# Near-Earth_object

A **near-Earth object** (**NEO**) is a Solar System object whose orbit brings it into proximity with the Earth. All NEOs have a closest approach to the Sun (perihelion) of less than 1.3 AU.[1] They include a few thousand near-Earth asteroids (NEAs), near-Earth comets, a number of solar-orbiting spacecraft, and meteoroids large enough to be tracked in space before striking the Earth. It is now widely accepted that collisions in the past have had a significant role in shaping the geological and biological history of the planet.[2] NEOs have become of increased interest since the 1980s because of increased awareness of the potential danger some of the asteroids or comets pose to the Earth, and active mitigations are being researched.

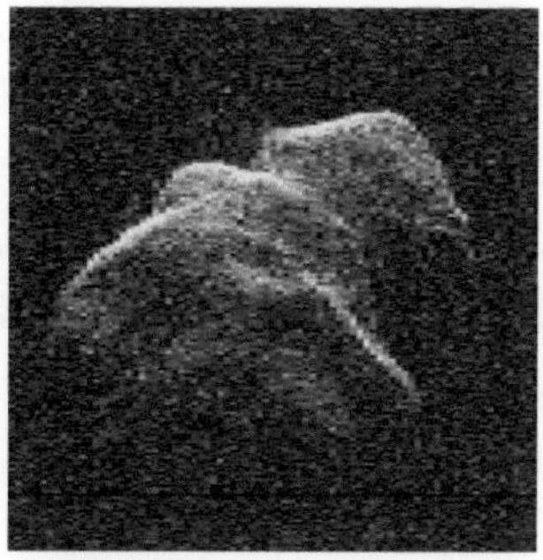

Asteroid 4179 Toutatis is a potentially hazardous object that has passed within 2.3 lunar distances.

Those NEOs that are asteroids (NEA) have orbits that lie partly between 0.983 and 1.3 astronomical units away from the Sun.[3] When an NEA is detected it is submitted to the IAU's Minor Planet Center (located at the Harvard-Smithsonian Center for Astrophysics) for cataloging. Some near-Earth asteroids' orbits intersect that of Earth's so they pose a collision danger.[4] The United States, European Union and other nations are currently scanning for NEOs[5] in an effort called Spaceguard.

Asteroid Toutatis from Paranal.

In the United States, NASA has a congressional mandate to catalogue all NEOs that are at least 1 kilometer wide, as the impact of such an object would be catastrophic. As of August 2012, there had been 848 near-Earth asteroids larger than 1km discovered, but only 154 are potentially hazardous asteroids (PHAs).[6] It was estimated in 2006 that 20% of the mandated objects have not yet been found.[5] As a result of NEOWISE in 2011, it is estimated that 93% of the NEAs larger than 1km have been found and that only about 70 remain to be discovered.[7]

Potentially hazardous objects (PHOs) are currently defined based on parameters that measure the object's potential to make threatening close approaches to the Earth.[8] Mostly objects with an Earth minimum orbit intersection distance (MOID) of 0.05 AU or less and an absolute magnitude (H) of 22.0 or less (a rough indicator of large size) are considered PHOs. Objects that cannot approach closer to the Earth (i.e. MOID) than 0.05 AU ( km; mi), or are smaller than about 150 m (500 ft) in diameter (i.e. H = 22.0 with assumed albedo of 13%), are not considered PHOs.[1] The NASA Near Earth Object Catalog also includes the approach distances of asteroids and comets measured in lunar distances,[9] and this usage has become the more usual unit of measure used by the press and mainstream media in discussing these objects.

Some NEOs are of high interest because they can be physically explored with lower mission velocity even than the Moon, due to their combination of low velocity with respect to Earth ($\Delta V$) and small gravity, so they may present interesting scientific opportunities both for direct geochemical and astronomical investigation, and as potentially economical sources of extraterrestrial materials for human exploitation.[10] This makes them an attractive target for exploration.[11] As of 2012, three near-Earth objects have been visited by spacecraft: 433 Eros, by NASA's Near Earth Asteroid Rendezvous probe,[12] 25143 Itokawa, by the JAXA Hayabusa mission,[13] and 4179 Toutatis, by

CNSA's Chang'e 2 spacecraft.[14]

## Risk

The risk that any near-Earth object poses is typically seen through a lens that is a function of both the culture and the technology of human society. "NEOs have been understood differently throughout history." Each time an NEO is observed, "a different risk was posed, and throughout time that risk perception has evolved. It is not just a matter of scientific knowledge."[15]

Such perception of risk is thus "a product of religious belief, philosophic principles, scientific understanding, technological capabilities, and even economical resourcefulness."[15]

### Risk scales

There are two schemes for classification of impact hazards from NEOs:

- the simple Torino Scale, and
- the more complex Palermo Technical Impact Hazard Scale

The annual background frequency used in the Palermo scale for impacts of energy greater than *E* megatonnes is estimated as:[16]

For instance, this formula implies that the expected value of the time from now until the next impact greater than 1 megatonne is 33 years, and that when it occurs, there is a 50% chance that it will be above 2.4 megatonnes. This formula is only valid over a certain range of *E*.

However, another paper[17] published in 2002 – the same year as the paper on which the Palermo scale is based – found a power law with different constants:

This formula gives considerably lower rates for a given *E*. For instance, it gives the rate for bolides of 10 megatonnes or more (like the Tunguska explosion) as 1 per thousand years, rather than 1 per 210 years as in the Palermo formula. However, the authors give a rather large uncertainty (once in 400 to 1800 years for 10 megatonnes), due in part to uncertainties in determining the energies of the atmospheric impacts that they used in their determination.

### Highly rated risks

On 25 December 2004, minor planet 99942 Apophis (at the time known by its provisional designation 2004 $MN_4$) was assigned a 4 on the Torino scale, the highest rating so far. On 27 December 2004, there was a 2.7% chance of Earth impact on 13 April 2029. However, on 28 December 2004, the risk of impact dropped to zero for 2029, but, due to a resonant return possibility the Torino rating for an April 2036 impact rose to 4 in early 2005, but by October 2009 the Torino rating was 0 (zero). The Palermo rating (October 2009) is −3.08.[18]

As of 12 September 2011, the only known NEO with a Palermo scale value greater than zero is (29075) 1950 DA, which is predicted to pass very close to or collide with the Earth ($p \leq 0.003$) in the year 2880. Depending on the orientation of its axis of rotation, it will either miss the Earth by tens of millions of kilometers, or have a 1 in 300 chance of hitting the Earth. However, humanity has over 800 years to refine its estimates of the orbit of (29075) 1950 DA, and to deflect it, if necessary.[19]

### List of current threats

NASA maintains a continuously updated web page of the most significant NEO threats in the next 100 years.[18] All or nearly all of the items on this page are highly likely to drop off the list eventually as more observations come in, reducing the uncertainties and enabling more accurate orbital predictions. (The page does not include 1950 DA, because that will not strike for at least 800 years.)[19]

## Number and classification of near-Earth objects

While orbiting the Sun, most potential impactors can be classified as meteoroids, asteroids, or comets depending on size and composition. Asteroids can also be members of an asteroid family, and comets can leave debris in their orbits.

As of January 2013, 9,534 NEOs have been discovered:[6] 93 near-Earth comets and 9,441 near-Earth Asteroids. Of those there are 742 Aten asteroids, 3,569 Amor asteroids, and 5,118 Apollo asteroids. There are 1,360 NEOs that are classified as potentially hazardous asteroids (PHAs). Currently, 154 PHAs and 858 NEAs have an absolute magnitude of 17.75 or brighter, which roughly corresponds to at least 1 km in size.[6]

As of January 2013, there are 427 NEAs on the Sentry impact risk page at the NASA website.[18] A significant number of these NEAs – 215 as of May 2010 – are equal to or smaller than 50 meters in diameter and none of the listed objects are placed even in the "yellow zone" (Torino Scale 2), meaning that none warrant the attention of general public.[20] As of January 2013, only asteroid 2007 $VK_{184}$ is listed as having a Torino Scale of 1.

### Near-Earth meteoroids

Near-Earth meteoroids are smaller near-Earth asteroids having an estimated diameter less than 50 meters. They are listed as asteroids on most asteroid tables. The JPL Small-Body Database lists 1,349 near Earth asteroids with an absolute magnitude (H) dimmer than 25 (roughly 50 meters in diameter).[21] The smallest known near-Earth meteoroid is 2008 $TS_{26}$ with an absolute magnitude of 33 and estimated size of only 1 meter.[21]

### Near-Earth asteroids

These are objects in a near-Earth orbit without the tail or coma of a comet. As of May 2012, 8,880 near-Earth asteroids are known,[6] ranging in size from 1 meter up to ~32 kilometers (1036 Ganymed). The number of near-Earth asteroids over one kilometer in diameter is estimated to be about 981.[7] [22] [23] The composition of near-Earth asteroids is comparable to that of asteroids from the asteroid belt, reflecting a variety of asteroid spectral types.[24]

There are significantly fewer near-Earth asteroids in the mid-size range than previously thought.

NEAs survive in their orbits for just a few million years.[3] They are eventually eliminated by planetary perturbations which cause ejection from the Solar System or a collision with the Sun or a planet. With orbital lifetimes short compared to the age of the Solar System, new asteroids must be constantly moved into near-Earth orbits to explain the observed asteroids. The accepted origin of these asteroids is that asteroid-belt asteroids are moved into the inner Solar System through orbital resonances with Jupiter. The interaction with Jupiter through the resonance perturbs the asteroid's orbit and it comes into the inner Solar System. The asteroid belt has gaps, known as Kirkwood gaps, where these resonances occur as the asteroids in these resonances have been moved onto other orbits. New asteroids migrate into these resonances, due to the Yarkovsky effect that provides a continuing supply of near-Earth asteroids.[25]

A small number of NEOs are extinct comets that have lost their volatile surface materials, although having a faint or intermittent comet-like tail does not necessarily result in a classification as a near-Earth comet, making the boundaries somewhat fuzzy. The rest of the near-Earth asteroids are driven out of the asteroid belt by gravitational interactions with Jupiter.[3] [26]

Near-Earth asteroids are divided into groups based on their semi-major axis (a), perihelion distance (q), and aphelion distance (Q)[1] [3] :

- The *Atiras* or *Apohele asteroids* have orbits strictly inside Earth's orbit: an Atira asteroid's aphelion distance (Q) is smaller than Earth's perihelion distance (0.983 AU). That is, Q < 0.983 AU. (This implies that the asteroid's semi-major axis is also less than 0.983 AU.)
- The *Atens* have a semi-major axis of less than 1 AU and cross Earth's orbit. Mathematically, a < 1.0 AU and Q > 0.983 AU.
- The *Apollos* have a semi-major axis of more than 1 AU and cross Earth's orbit. Mathematically, a > 1.0 AU and q < 1.017 AU. (1.017 AU is Earth's aphelion distance.)
- The *Amors* have orbits strictly outside Earth's orbit: an Amor asteroid's perihelion distance (q) is greater than Earth's aphelion distance (1.017 AU). Amor asteroids are also near-earth objects so q < 1.3 AU. In summary, 1.017 AU < q < 1.3 AU. (This implies that the asteroid's semi-major axis (a) is also larger than 1.017 AU.) Some Amor asteroid orbits cross the orbit of Mars.

(Note: Some authors define the Atens group differently: they define it as being all the asteroids with a semi-major axis of less than 1 AU. That is, they consider the Atiras to be part of the Atens. Historically, until 1998, there were no known or suspected Atiras, so the distinction wasn't necessary.)

Because the Atens and all Apollos have orbits that cross Earth's orbit, they might impact the Earth. Atiras and Amors do not cross the Earth's orbit and are not immediate impact threats, but their orbits may change to become Earth-crossing orbits in the future.

Also sometimes used is the *Arjuna asteroid* classification, for asteroids with extremely Earth-like orbits.[27]

### Near-Earth comets

As of May 2012, 91 near-Earth comets have been discovered.[6] Although no impact of a comet in Earth´s history has been conclusively confirmed, the Tunguska event may have been caused by a fragment of Comet Encke.[28] Cometary fragmenting may also be responsible for some impacts from near-Earth objects. It is rare for a comet to pass within 0.1 AU ( km; mi) of Earth.[29]

These near-Earth objects were probably derived from the Kuiper belt, beyond the orbit of Neptune.

## Impact rate

Objects with diameters of 5-10 m impact the Earth's atmosphere approximately once per year, with as much energy as the atomic bomb dropped on Hiroshima, approximately 15 kilotonnes of TNT. These ordinarily explode in the upper atmosphere, and most or all of the solids are vaporized.[30] Every 2000–3000 years NEAs produce explosions comparable to the one observed at Tunguska in 1908.[31] Objects with a diameter of one kilometer hit the Earth an average of twice every million year interval.[3] Large collisions with five kilometer objects happen approximately once every ten million years.

Assuming that these rates will continue for the next billion years, there exist at least 2,000 objects of diameter greater than 1 km that will eventually hit Earth. However, most of these are not yet considered potentially hazardous objects because they are currently orbiting between Mars and Jupiter. Eventually they will change orbits and become NEOs. Objects spend on average a few million years as NEOs before hitting the Sun, being ejected from the Solar System, or (for a small proportion) hitting a planet.[32]

## Historic impacts

The general acceptance of the Alvarez hypothesis, explaining the Cretaceous–Paleogene extinction event as the result of a large object impact event, raised the awareness of the possibility of future Earth impacts with other objects that cross the Earth's orbit.[2]

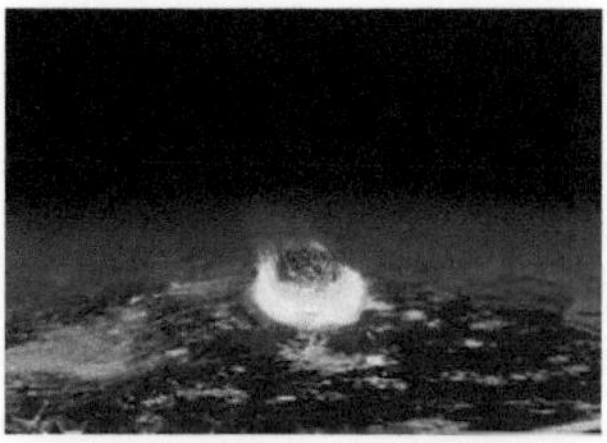

Illustration of the impact of an asteroid a few kilometers across. Such impacts are expected to occur less often than every 100 million years.

### 1908 Tunguska event

It is now commonly believed that on 30 June 1908 a stony asteroid exploded over Tunguska with the energy of the explosion of 10 megatons of TNT. The explosion occurred at a height of 8.5 kilometers. The object that caused the explosion has been estimated to have had a diameter of 45–70 meters.[33]

### 1979 Vela Incident

A 22 September 1979 event recorded as occurring near the junction of the South Atlantic and the Indian Ocean was possibly a low-yield nuclear test, but was also initially thought to have been caused by the possible impact of an extraterrestrial object. The event, which became known as the Vela Incident, was identified by a U.S. Vela defense satellite in Earth orbit. The event alarm triggered multi-year investigations by several organizations which could not conclusively determine if the explosion was of nuclear or non-nuclear origin.

### 2002 Eastern Mediterranean event

On 6 June 2002 an object with an estimated diameter of 10 meters collided with Earth. The collision occurred over the Mediterranean Sea, between Greece and Libya, at approximately 34°N 21°E and the object exploded in mid-air. The energy released was estimated (from infrasound measurements) to be equivalent to 26 kilotons of TNT, comparable to a small nuclear weapon.[34]

### 2008 Sudan event

On 6 October 2008, scientists calculated that a small near-Earth asteroid, 2008 $TC_3$, just sighted that night, would impact the Earth on 7 October over Sudan, at 0246 UTC, 5:46 local time.[35] [36] The asteroid arrived as predicted.[37] [38] This is the first time that an asteroid impact on Earth has been accurately predicted. While hundreds of rock fragments have subsequently been recovered on ground,[39] no reports of the actual impact have so far been published since it occurred in a very sparsely populated area.[40]

### 2009 Indonesia event

A large fireball was observed in the skies near Bone, Indonesia on October 8, 2009. This was thought to be caused by an asteroid approximately 10 meters in diameter. The fireball contained an estimated energy of 50 kilotons of TNT, or about twice the Nagasaki atomic bomb. No injuries were reported.[41]

## Close approaches

Flyby of asteroid 2004 FH. The other object that flashes by is an artificial satellite.

On August 10, 1972 a meteor that became known as The Great Daylight 1972 Fireball was witnessed by many people moving north over the Rocky Mountains from the U.S. Southwest to Canada. It was an Earth-grazing meteoroid that passed within 57 kilometers (about 34 miles) of the Earth's surface. It was filmed by a tourist at the Grand Teton National Park in Wyoming with an 8-millimeter color movie camera.[42]

On March 23, 1989 the 300-meter (1,000-foot) diameter Apollo asteroid 4581 Asclepius (1989 FC) missed the Earth by **unknown operator: u','** kilometers (**unknown operator: u'strong'** mi) passing through the exact position where the Earth was only 6 hours before. If the asteroid had impacted it would have created the largest explosion in recorded history, 12 times as powerful as the Tsar Bomba, the most powerful nuclear bomb ever exploded. It attracted widespread attention as early calculations had its passage being as close as **unknown operator: u','unknown operator: u','unknown operator: u','** (**unknown operator: u'strong'** mi) from the Earth, with large uncertainties that allowed for the possibility of it striking the Earth.[43]

On March 18, 2004, LINEAR announced a 30-meter asteroid, 2004 FH, which would pass the Earth that day at only **unknown operator: u','unknown operator: u','unknown operator: u','** (**unknown operator: u'strong'** mi), about one-tenth the distance to the Moon, and the closest miss ever noticed. They estimated that similar-sized asteroids come as close about every two years.[44]

On March 31, 2004, two weeks after 2004 FH, meteoroid 2004 $FU_{162}$ set a new record for closest recorded approach, passing Earth only **unknown operator: u','unknown operator: u','unknown operator: u','** (**unknown operator: u'strong'** mi) away (about one-sixtieth of the distance to the Moon). Because it was very small (6 meters/20 feet), $FU_{162}$ was detected only hours before its closest approach. If it had collided with Earth, it probably would have harmlessly disintegrated in the atmosphere.

On March 2, 2009, near-Earth asteroid 2009 $DD_{45}$ flew by Earth at about 13:40 UT. The estimated distance from Earth was **unknown operator: u','unknown operator: u','unknown operator: u','** (**unknown operator: u'strong'** mi), approximately twice the height of a geostationary communications satellite. The estimated size of the space rock was about 35 meters (115 feet) wide.[45]

On January 13, 2010 at 12:46 UT, near-Earth asteroid 2010 $AL_{30}$[46] passed at about **unknown operator: u','unknown operator: u','unknown operator: u','** (**unknown operator: u'strong'** mi). It was approximately 10–15 m ( ) wide. If 2010 $AL_{30}$ had entered the Earth's atmosphere, it would have created an air burst equivalent to between 50 kT and 100 kT (kilotons of TNT). The Hiroshima "Little Boy" atom bomb had a yield between 13-18kT.[47]

On June 28, 2011 an asteroid designated 2011 MD, estimated at 5–20 m ( ) in diameter, passed within **unknown operator: u','unknown operator: u','unknown operator: u','** (**unknown operator: u'strong'** mi) of the Earth, passing over the Atlantic Ocean.[48]

On November 8, 2011 2005 $YU_{55}$ (at about 400m diameter) passed within 324600 km ( mi) (0.85 lunar distances) of Earth. Ten weeks later, on January 27, 2012, the 10-metre wide asteroid 2012 $BX_{34}$ passed a mere 60000 km ( mi) from Earth.[49]

On February 15, 2013, 2012 $DA_{14}$ will pass approximately 0.00023 AU ( km; mi) from the center-point of Earth. This is potentially closer than satellites in geosynchronous orbit. The asteroid will not be visible to the naked eye.

## Future impacts

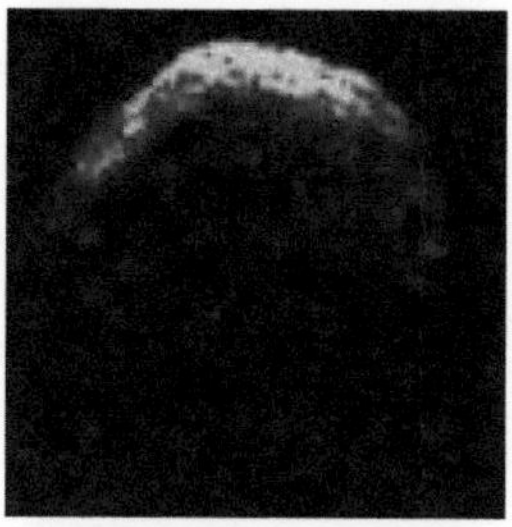

Radar image of asteroid 1950 DA.

Although there have been a few false alarms, a number of objects have been known to be threats to the Earth. (89959) 2002 NT7 was the first asteroid with a positive rating on the Palermo Technical Impact Hazard Scale, with approximately one in a million on a potential impact date of February 1, 2019; it is now known that on January 13, 2019, 2002 $NT_7$ will safely pass 0.4078 AU ( km; mi) from the Earth.[50]

Asteroid (29075) 1950 DA was lost after its discovery in 1950, since not enough observations were made to allow plotting of its orbit, and then rediscovered on December 31, 2000. The chance it will impact Earth on March 16, 2880, during its close approach has been estimated as 1 in 300. This chance of impact for such a large object is roughly 50% greater than that for all other such objects combined through 2880.[51] It has a diameter of about a kilometer (0.6 miles).

Only the asteroids 99942 Apophis (provisionally known as 2004 MN4) and (144898) 2004 $VD_{17}$ have briefly had above-normal rankings on the Torino Scale.

## Projects to minimize the threat

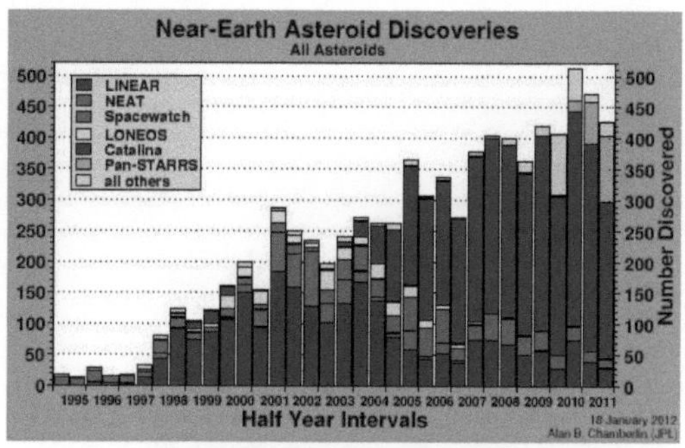


Number of NEOs detected by various projects.

Several surveys have undertaken "Spaceguard" activities (an umbrella term), including Lincoln Near-Earth Asteroid Research (LINEAR), Spacewatch, Near-Earth Asteroid Tracking (NEAT), Lowell Observatory Near-Earth-Object Search (LONEOS), Catalina Sky Survey, Campo Imperatore Near-Earth Objects Survey (CINEOS), Japanese Spaceguard Association, and Asiago-DLR Asteroid Survey. In 1998, the United States Congress mandated the Spaceguard Survey – detection of 90% of near-earth asteroids over 1 km diameter (which threaten global devastation) by 2008. This could be extended by the George E. Brown, Jr. Near-Earth Object Survey Act, which calls for NASA to detect 90 percent of NEOs with diameters of 140 meters or greater by 2020.[52]

As of 2011, 911 of the largest (>1 km diameter) near-Earth asteroids have been found, with an estimate of 70 yet to be found.[7]

On April 25, 2012, the Ecuadorian Space Agency announced that due a technological change in the video camera of its first satellite it will be adding a 3rd mission, which is to help organizations and individuals around the world monitor for small, most dangerous near-earth objects directly from orbit in real time and help fine tune the catalog of orbital debris by injecting the live video signal of the satellite into the Internet, thus turning the NEE-01 Pegasus in the first online, real-time video orbital sentry.[53]

## See also

- List of Earth-crossing minor planets
- 6Q0B44E, in Earth orbit with a period of 80 days
- J002E3, probably the third stage of the Apollo 12 Saturn V
- BRAMS (Belgian RAdio Meteor Stations); study of meteor population[54]
- Bolide
- Co-orbital configuration
- Euronear
- Mission Marco Polo
- NEODyS
- Orbit@home
- Space debris
- 2010 $SO_{16}$
- Asteroid mining
- Potentially hazardous object
- NEE-01 Pegasus

## References

[1] "NEO Groups" (http://neo.jpl.nasa.gov/neo/groups.html). NASA/JPL Near-Earth Object Program Office. . Retrieved 2012-06-04.

[2] Richard Monastersky (March 1, 1997). "The Call of Catastrophes" (http://www.sciencenews.org/pages/sn_arc97/75th/rm_essay.htm). Science News Online. . Retrieved 2007-10-23.

[3] A. Morbidelli, W. F. Bottke Jr., Ch. Froeschlé, P. Michel (January 2002). W. F. Bottke Jr., A. Cellino, P. Paolicchi, and R. P. Binzel. ed. "Origin and Evolution of Near-Earth Objects" (http://www.boulder.swri.edu/~bottke/Reprints/Morbidelli-etal_2002_AstIII_NEOs.pdf) (PDF). *Asteroids III* (University of Arizona Press): 409–422. .

[4] Clark R. Chapman (May 2004). "The hazard of near-Earth asteroid impacts on earth". *Earth and Planetary Science Letters* **222** (1): 1–15. Bibcode 2004E&PSL.222....1C. doi:10.1016/j.epsl.2004.03.004.

[5] Shiga, David (2006-06-27). "New telescope will hunt dangerous asteroids" (http://www.newscientist.com/article/dn9403). New Scientist. . Retrieved 2008-11-15.

[6] "NEO Discovery Statistics" (http://neo.jpl.nasa.gov/stats/). . Retrieved 2012-05-17.

[7] "WISE Revises Numbers of Asteroids Near Earth" (http://www.nasa.gov/mission_pages/WISE/multimedia/gallery/neowise/pia14734.html). NASA/JPL. 09.29.11. . Retrieved 2012-05-17. (NASA Space Telescope Finds Fewer Asteroids Near Earth) (http://www.nasa.gov/mission_pages/WISE/news/wise20110929.html)

[8] "Potentially Hazard Asteroids" (http://neo.jpl.nasa.gov/neo/pha.html). NASA/JPL Near-Earth Object Program Office. . Retrieved 2011-05-05.

[9] NEO Earth Close Approaches (http://neo.jpl.nasa.gov/ca/) at NASA/JPL Near-Earth Object Program Office

[10] Dan Vergano (February 2, 2007). "Near-Earth asteroids could be 'steppingstones to Mars'" (http://www.usatoday.com/tech/science/space/2007-02-12-asteroid_x.htm). USA Today. . Retrieved 2007-10-22.

[11] Rui Xu, Pingyuan Cui, Dong Qiao and Enjie Luan (18 March 2007). "Design and optimization of trajectory to Near-Earth asteroid for sample return mission using gravity assists". *Advances in Space Research* **40** (2): 200–225. Bibcode 2007AdSpR..40..220X. doi:10.1016/j.asr.2007.03.025.

[12] Donald Savage and Michael Buckley (January 31, 2001). "NEAR Mission Completes Main Task, Now Will Go Where No Spacecraft Has Gone Before" (http://nssdc.gsfc.nasa.gov/planetary/news/near_descent_pr_20010131.html). National Aeronautics and Space Administration. . Retrieved 2007-10-22.

[13] Don Yeomans (August 11, 2005). "Hayabusa's Contributions Toward Understanding the Earth's Neighborhood" (http://neo.jpl.nasa.gov/missions/hayabusa.html). National Aeronautics and Space Administration. . Retrieved 2007-10-22.

[14] Emily Lakdawalla. "Chang'e 2 imaging of Toutatis" (http://www.planetary.org/blogs/emily-lakdawalla/2012/12141551-change-2-imaging-of-toutatis.html). .

[15] Fernández Carril, Luis (2012-05-14). "The evolution of near Earth objects risk perception" (http://www.thespacereview.com/article/2080/1). *The Space Review*. . Retrieved 2012-05-25.

[16] "THE PALERMO TECHNICAL IMPACT HAZARD SCALE" (http://neo.jpl.nasa.gov/risk/doc/palermo.html). NASA/JPL Near-Earth Object Program Office. 31 Aug 2005. . Retrieved 2008-07-27.

[17] "The flux of small near-Earth objects colliding with the Earth" by P. Brown et al., Nature, 420, pp. 294–6, November 2002.

[18] "Current Impact Risks" (http://neo.jpl.nasa.gov/risk). . Retrieved 2010-05-28.

[19] "Asteroid 1950 DA" (http://neo.jpl.nasa.gov/1950da/). NASA/JPL Near-Earth Object Program Office. . Retrieved 2011-10-14.

[20] The Torino Impact Hazard Scale (http://neo.jpl.nasa.gov/torino_scale.html)

[21] "JPL Small-Body Database Search Engine: asteroids and NEOs and H > 25 (mag)" (http://ssd.jpl.nasa.gov/sbdb_query.cgi?obj_group=neo;obj_kind=ast;obj_numbered=all;OBJ_field=0;ORB_field=0;c1_group=OBJ;c1_item=Ai;c1_op=>;c1_value=25;table_format=HTML;max_ cgifields=format_option;.cgifields=ast_orbit_class;.cgifields=table_format;.cgifields=obj_kind;.cgifields=obj_group;.cgifields=obj_numbered;.cgifields=com_orbit_class&query=1&c_sort=AiD). JPL Small-Body Database. . Retrieved 2012-05-18.

[22] Jane Platt (January 12, 2000). "Asteroid Population Count Slashed" (http://www.jpl.nasa.gov/releases/2000/neat.html). National Aeronautics and Space Administration. . Retrieved 2007-10-22.

[23] David Rabinowitz, Eleanor Helin, Kenneth Lawrence and Steven Pravdo (13 January 2000). "A reduced estimate of the number of kilometer-sized near-Earth asteroids" (http://www.nature.com/nature/journal/v403/n6766/full/403165a0.html). *Nature* **403** (6766): 165–166. Bibcode 2000Natur.403..165R. doi:10.1038/35003128. PMID 10646594. . Retrieved 2007-10-22.

[24] D.F. Lupishko and T.A. Lupishko (May 2001). "On the Origins of Earth-Approaching Asteroids". *Solar System Research* **35** (3): 227–233. Bibcode 2001SoSyR..35..227L. doi:10.1023/A:1010431023010.

[25] A. Morbidelli, D. Vokrouhlický (May 2003). "The Yarkovsky-driven origin of near-Earth asteroids". *Icarus* **163** (1): 120–134. Bibcode 2003Icar..163..120M. doi:10.1016/S0019-1035(03)00047-2.

[26] D.F. Lupishko, M. di Martino and T.A. Lupishko (September 2000). "What the physical properties of near-Earth asteroids tell us about sources of their origin?". *Kinematika i Fizika Nebesnykh Tel Supplimen* **3** (3): 213–216. Bibcode 2000KFNTS...3..213L.

[27] Ron Cowen (February 20, 1993). "Near-Earth asteroids: class consciousness – new asteroids identified" (http://findarticles.com/p/articles/mi_m1200/is_n8_v143/ai_13526583). Science News. . Retrieved 2007-10-23.

[28] "The Tunguska object – A fragment of Comet Encke" (http://adsabs.harvard.edu/abs/1978BAICz..29..129K). Astronomical Institutes of Czechoslovakia. . Retrieved 2007-10-15.

[29] "Closest Approaches to the Earth by Comets" (http://www.minorplanetcenter.org/iau/lists/ClosestComets.html). Minor Planet Center. . Retrieved 2012-06-28.

[30] Clark R. Chapman & David Morrison (6 January 1994). "Impacts on the Earth by asteroids and comets: assessing the hazard". *Nature* **367** (6458): 33–40. Bibcode 1994Natur.367...33C. doi:10.1038/367033a0.

[31] EARTH IN THE COSMIC SHOOTING GALLERY, D. J. Asher et al. page 2 (http://www.arm.ac.uk/preprints/455.pdf)

[32] Morbidelli, A., W. F. Bottke, Ch. Froeschle, and P. Michel. 2002. Origin and evolution of near-Earth objects. (http://www.boulder.swri.edu/~bottke/Reprints/Morbidelli-etal_2002_AstIII_NEOs.pdf) In Asteroids III, (W. F. Bottke, A. Cellino, P. Paolicchi, R. Binzel, Eds). U. Arizona Press, 409-422.

[33] Christopher F. Chyba, Paul J. Thomas & Kevin J. Zahnle (January 7, 1993). "The 1908 Tunguska explosion: atmospheric disruption of a stony asteroid". *Nature* **361** (6407): 40–44. Bibcode 1993Natur.361...40C. doi:10.1038/361040a0.

[34] P. Brown, R.E. Spalding, D.O. ReVelle, E. Tagliaferri and S.P. Worden (21 November 2002). "The flux of small near-Earth objects colliding with the [[Earth (http://www.astro.uwo.ca/~pbrown/documents/flux-final.pdf)]"] (PDF). *Nature* **420** (6913): 294–296. doi:10.1038/nature01238. PMID 12447433. . Retrieved 2007-10-23.

[35] Don Yeomans (October 6, 2008). "Small Asteroid Predicted to Cause Brilliant Fireball over Northern Sudan" (http://neo.jpl.nasa.gov/news/news159.html). NASA/JPL Near-Earth Object Program Office. . Retrieved 2008-10-09.

[36] Richard A. Kerr (6 October 2008). "FLASH! Meteor to Explode Tonight" (http://sciencenow.sciencemag.org/cgi/content/full/2008/1006/2). ScienceNOW Daily News. . Retrieved 2008-10-09.

[37] Don Yeomans (October 7, 2008). "Impact of Asteroid 2008 TC3 Confirmed" (http://neo.jpl.nasa.gov/news/news160.html). NASA/JPL Near-Earth Object Program Office. . Retrieved 2008-10-09.

[38] Richard A. Kerr (8 October 2008). "Asteroid Watchers Score a Hit" (http://sciencenow.sciencemag.org/cgi/content/full/2008/1008/1). ScienceNOW Daily News. . Retrieved 2008-10-09.

[39] P. Jenniskens et. al. (2009-03-26). "The impact and recovery of asteroid 2008 TC3" (http://www.nature.com/nature/journal/v458/n7237/abs/nature07920.html). *Nature* **458** (7237): 485–488. Bibcode 2009Natur.458..485J. doi:10.1038/nature07920. PMID 19325630. Archived (http://www.webcitation.org/5hDZOG4BF) from the original on 2009-06-01. . Retrieved 2009-04-04. Published in Letters to Nature

[40] Little Asteroid Makes a Big Splash (http://www.skyandtelescope.com/community/skyblog/newsblog/30686199.html) Sky and Telescope, October 9, 2008.

[41] Don Yeomans, Paul Chodas and Steve Chesley (October 23, 2009). "Asteroid Impactor Reported over Indonesia" (http://neo.jpl.nasa.gov/news/news165.html). NASA/JPL Near-Earth Object Program Office. . Retrieved 2009-10-30.

[42] Grand Teton Meteor Video (http://www.youtube.com/watch?v=7M8LQ7_hWtE), Youtube

[43] Brian G. Marsden (1998-03-29). "HOW THE ASTEROID STORY HIT: AN ASTRONOMER REVEALS HOW A DISCOVERY SPUN OUT OF CONTROL" (http://www.cfa.harvard.edu/iau/pressinfo/1997XF11Globe.html). The Boston Globe. . Retrieved 2007-10-23.

[44] Steven R. Chesley and Paul W. Chodas (March 17, 2004). "Recently Discovered Near-Earth Asteroid Makes Record-breaking Approach to Earth" (http://neo.jpl.nasa.gov/news/news142.html). National Aeronautics and Space Administration. . Retrieved 2007-10-23.

[45] Yahoo News – 20090302 – Asteroid Flies Past Earth (http://news.yahoo.com/s/space/20090302/sc_space/asteroidfliespastearth)
[46] Small Asteroid 2010 AL30 Will Fly Past The Earth (http://neo.jpl.nasa.gov/news/news167.html). NASA/JPL Near-Earth Object Program, January 12, 2010.
[47] Near-Earth Object 2010 AL30 (http://epod.usra.edu/blog/2010/03/nearearth-object-2010-al30.html). NASA Earth Science Picture of the Day March 06, 2010.
[48] "Asteroid soars over Atlantic Ocean" (http://www.abc.net.au/news/stories/2011/06/28/3255543.htm). Australian Broadcasting Corporation. June 28, 2011. . Retrieved 2011-07-03.
[49] "Asteroid makes near-miss fly-by" (http://www.bbc.co.uk/news/science-environment-16756450). British Broadcasting Corporation. 27 January 2012. . Retrieved 2012-01-28.
[50] "JPL Close-Approach Data: 89959 (2002 NT7)" (http://ssd.jpl.nasa.gov/sbdb.cgi?sstr=2002NT7;cad=1#cad). 2011-09-12 last obs (arc=57 years). . Retrieved 2011-11-04.
[51] Giorgini, J. D.; Ostro, S. J.; Benner, L. A. M.; Chodas, P. W.; Chesley, S. R.; Hudson, R. S.; Nolan, M. C.; Klemola, A. R. et al. (April 5, 2002). "Asteroid 1950 DA's Encounter with Earth in 2880: Physical Limits of Collision Probability Prediction" (http://neo.jpl.nasa.gov/1950da/1950da.pdf) (PDF). *Science* **296** (5565): 132–136. Bibcode 2002Sci...296..132G. doi:10.1126/science.1068191. PMID 11935024. . Retrieved January 11, 2010.
[52] National Academy of Sciences Jan. 23, 2010 (http://www8.nationalacademies.org/onpinews/newsitem.aspx?RecordID=12842) Defending Planet Earth: Near-Earth Object Surveys and Hazard Mitigation Strategies: Final Report. Washington, DC: The National Academies Press. Book available at: http://books.nap.edu/catalog.php?record_id=12842.
[53] http://exa.ec/bp42/index-en.html
[54] Belgian RAdio Meteor Stations (http://brams.aeronomie.be/)

# External links

- Interactive Earth damage impact calculator from NEO (http://www.purdue.edu/impactearth)
- The NEO Confirmation Page (http://www.minorplanetcenter.net/iau/NEO/ToConfirm.html) (NEOCP)
- The Jet Propulsion Laboratory's NEO website (http://neo.jpl.nasa.gov) / NEO Earth Close Approaches (http://neo.jpl.nasa.gov/ca/)
- Observable Near-Earth Asteroids (http://www.lowell.edu/LARI_display_asteroid_targets.php) (Lowell Observatory)
- TECA Table of next close approaches to the Earth (http://www.brera.mi.astro.it/sormano/teca.html)
- SAEL Small Asteroids Encounter List (http://www.brera.mi.astro.it/sormano/sael.html)
- MBPL Minor Body Priority List (http://www.brera.mi.astro.it/sormano/mbpl.html)
- Spaceguard UK (http://www.spaceguarduk.com/)
- House Holds Hearing On NEO Threat (http://starrymirror.com/neo hearing1109.htm)
- The UK NEO Information Centre (http://www.spacecentre.co.uk/Page.aspx/6/NEAR_EARTH_OBJECTS/)
- Earth in the Cosmic Shooting Gallery (http://www.arm.ac.uk/preprints/455.pdf) Paper published 2005
- Near Earth Asteroids (http://www.astronomycast.com/astronomy/episode-29-asteroids-make-bad-neighbors/) Astronomy Cast episode #29, includes full transcript.
- Upcoming Close Approaches (<0.10 A.U.) of Near-Earth Objects to Earth (http://members.shaw.ca/andrewlowe/PHA-SOON.HTM#Earth)
- Catalogue of the Solar System Small Bodies Orbital Evolution (http://smallbodies.ru/en/)
- NASA Space Telescope Finds Fewer Asteroids Near Earth (http://www.nasa.gov/mission_pages/WISE/news/wise20110929.html) (09.29.11)

# Centaur_(minor_planet)

**Centaurs** are small Solar System bodies with a semi-major axis between those of the giant planets. They therefore have unstable orbits that cross or have crossed the orbits of one or more of the giant planets, and have dynamic lifetimes of a few million years.[1] Centaurs typically behave with characteristics of both asteroids and comets. They are named after the mythological race of beings, centaurs, which were a mixture of horse and human. It has been estimated that there are around 44,000 centaurs in the Solar System with diameters larger than 1 km.[1]

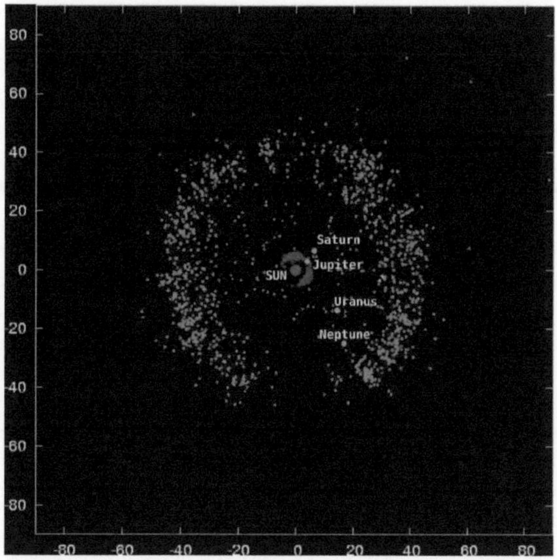

Positions of known outer Solar System objects. The centaurs are those objects (in orange) that lie generally **inwards** of the Kuiper belt (in green) and outside the Jupiter Trojans (pink).

The first centaur to be discovered was 944 Hidalgo in 1920. However, they were not recognized as a distinct population until the discovery of 2060 Chiron in 1977. The largest known centaur is 10199 Chariklo, discovered in 1997, which at 260 km in diameter is as big as a mid-sized main-belt asteroid.

No centaur has been photographed up close, although there is evidence that Saturn's moon Phoebe, imaged by the Cassini probe in 2004, may be a captured centaur. In addition, the Hubble Space Telescope has gleaned some information about the surface features of 8405 Asbolus.

As of 2008, three centaurs have been found to display cometary comas: Chiron, 60558 Echeclus, and 166P/NEAT. Chiron and Echeclus are therefore classified as both asteroids and comets. Other centaurs such as 52872 Okyrhoe are suspected of showing cometary activity. Any centaur that is perturbed close enough to the Sun is expected to become a comet.

## Classification

The generic definition of a centaur is a small body that orbits the Sun between Jupiter and Neptune and crosses the orbits of one or more of the giant planets. Due to the inherent long-term instability of orbits in this region, even centaurs such as 2000 $GM_{137}$ and 2001 $XZ_{255}$, which do not currently cross the orbit of any planet, are in gradually changing orbits that will be perturbed until they start to cross the orbit of one or more of the giant planets.[1]

However, different institutions have different criteria for classifying borderline objects, based on particular values of their orbital elements:

- The Minor Planet Center (MPC) defines centaurs as having a perihelion beyond the orbit of Jupiter and a semi-major axis less than that of Neptune.[2]

- The Jet Propulsion Laboratory (JPL) similarly defines centaurs as having a semi-major axis, $a$, between those of Jupiter and Neptune (5.5 AU < $a$ < 30.1 AU).[3]
- In contrast, the Deep Ecliptic Survey (DES) defines centaurs using a dynamical classification scheme. These classifications are based on the simulated change in behavior of the present orbit when extended over 10 million years. The DES defines centaurs as non-resonant objects whose instantaneous (osculating) perihelia are less than the osculating semi-major axis of Neptune at any time during the simulation. This definition is intended to be synonymous with planet-crossing orbits and to suggest comparatively short lifetimes in the current orbit.[4]

The collection *The Solar System Beyond Neptune* (2008) uses the traditional definition of centaurs, limited to semi-major axes smaller than that of Neptune, classifying the objects on unstable orbits beyond this limit as members of the scattered disk.[5] Yet, other astronomers still prefer to define centaurs as objects that are non-resonant with a perihelion inside the orbit of Neptune that can be shown to likely cross the Hill sphere of a gas giant within the next 10 million years.[6] Thus centaurs can be thought of as inward scattered objects that interact more aggressively and scatter more quickly than typical scattered disc objects.

The JPL Small-Body Database lists 183 centaurs.[7] There are an additional 40 trans-Neptunian objects with a semi-major axis further than Neptune (a > 30.1 AU) and perihelion closer than the orbit of Uranus (q < 19 AU).[7] The Committee on Small Body Nomenclature of the International Astronomical Union has not formally weighed in on either side of the debate. Instead, it has adopted the following naming convention for such objects: befitting their centaur-like transitional orbits between TNOs and comets, "objects on unstable, non-resonant, giant-planet-crossing orbits with semimajor axes greater than Neptune's" are to be named for other hybrid and shape-shifting mythical creatures. Thus far, only the binary objects Ceto and Phorcys and Typhon and Echidna have been named according to the new policy.[8]

Other objects caught between these differences in classification methods include (44594) 1999 $OX_3$, which has a semi-major axis of 32 AU but crosses the orbits of both Uranus and Neptune. Among the inner centaurs, 2005 VD, with a perihelion distance very near Jupiter, is listed as a centaur by both JPL and DES.

Centaurs with measured diameters listed as possible dwarf planets according to Mike Brown's website include 10199 Chariklo, 2060 Chiron, and 54598 Bienor.[9]

## Orbits

### Distribution

The diagram at right illustrates the orbits of all known centaurs in relation to the orbits of the planets. For selected objects, the eccentricity of the orbits is represented by red segments (extending from perihelion to aphelion).

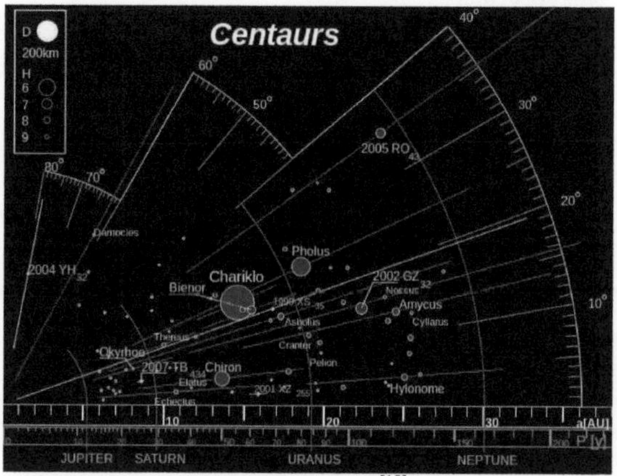

Orbits of known centaurs[10]

Centaurs' orbits are characterised by a wide range of eccentricity, from highly eccentric (Pholus, Asbolus, Amicus, Nessus) to more circular (Chariklo and the *Saturn-crossers*: Thereus, Okyrhoe).

To illustrate the range of the orbits' parameters, a few objects with very unusual orbits are plotted in yellow on the diagram:

- 1999 $XS_{35}$ (Apollo asteroid) follows an extremely eccentric orbit ($e$=0.947), leading it from inside Earth's orbit (0.94 AU) to well beyond Neptune (>34 AU)
- 2007 $TB_{434}$ follows a quasi-circular orbit ($e$<0.026)
- 2001 $XZ_{255}$ has the lowest inclination ($i$<3°).
- Damocles is among a few centaurs on orbits with extreme inclination (prograde $i$>70°, e.g. 2007 $DA_{61}$, 2004 $YH_{32}$, retrograde $i$<120° e.g. 2005 $JT_{50}$; not shown)
- 2004 $YH_{32}$ follows such a highly inclined orbit (nearly 80°) that, while it crosses from the distance of the asteroid belt from the Sun to past the distance of Saturn, it does not even cross Jupiter relative to the plane of Jupiter's orbit.

A dozen known centaurs, including Dioretsa ("asteroid" spelled backwards), follow retrograde orbits.

## Changing orbits

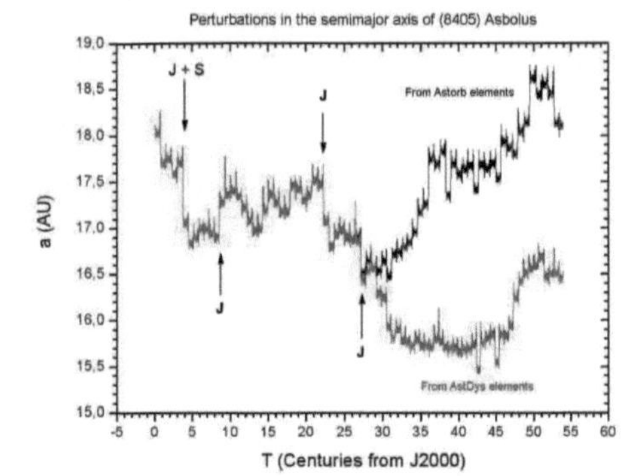


The semi-major axis of Asbolus during the next 5500 years. After the Jupiter encounter of year 4713 (27 Cy in the future) the two orbits become substantially divergent.[11]

Because the centaurs cross the orbits of the giant planets and are not protected by orbital resonances, their orbits are unstable within a timescale of $10^6$–$10^7$ years.[12] For example, 55576 Amycus is in an unstable orbit near the 3:4 resonance of Uranus.[1] Dynamical studies of their orbits indicate that centaurs are probably an intermediate orbital state of objects transitioning from the Kuiper belt to the Jupiter family of short-period comets. Objects may be perturbed from the Kuiper belt, whereupon they become Neptune-crossing and interact gravitationally with that planet (see theories of origin). They then become classed as centaurs, but their orbits are chaotic, evolving relatively rapidly as the centaur makes repeated close approaches to one or more of the outer planets. Some centaurs will evolve into Jupiter-crossing orbits whereupon their perihelia may become reduced into the inner Solar System and they may be reclassified as active comets in the Jupiter family if they display cometary activity. Centaurs will thus ultimately collide with the Sun or a planet or else they may be ejected into interstellar space after a close approach to one of the planets, particularly Jupiter.

# Physical characteristics

The relatively small size of centaurs precludes surface observations, but colour indices and spectra can indicate possible surface composition and can provide insight into the origin of the bodies.[12]

## Colours

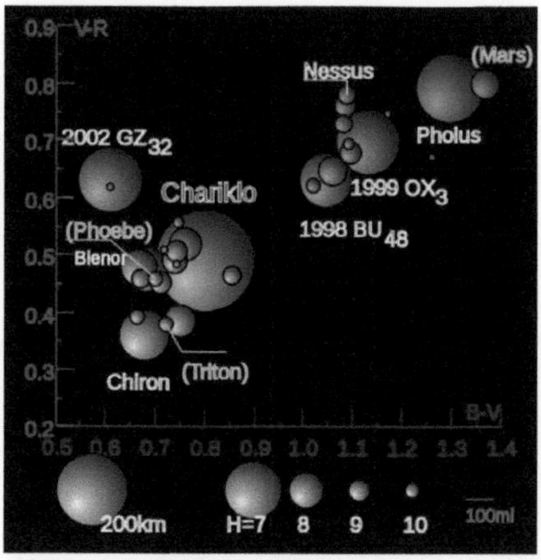

Colour distribution of centaurs

Centaurs display a puzzling diversity of colour that challenges any simple model of surface composition.[13] In the side-diagram, the colour indices are measures of apparent magnitude of an object through blue (B), visible (V) *i.e.* green-yellow and red (R) filters. The diagram illustrates these differences (in enhanced colour) for all centaurs with known colour indices. For reference, two moons: Triton and Phoebe, and planet Mars are plotted (yellow labels, size not to scale).

Centaurs appear to be grouped into two classes:

- very red, for example 5145 Pholus
- blue (or blue-grey, according to some authors), for example 2060 Chiron

There are numerous theories to explain this colour difference, but they can be divided broadly into two categories:

- The colour difference results from a difference in the origin and/or composition of the centaur (see origin below)
- The colour difference reflects a different level of space-weathering from radiation and/or cometary activity.

As examples of the second category, the reddish colour of Pholus has been explained as a possible mantle of irradiated red organics, whereas Chiron has instead had its ice exposed due to its periodic cometary activity, giving it a blue/grey index. The correlation with activity and color is not certain, however, as the active centaurs span the range of colors from blue (Chiron) to red (166P/NEAT).[14] Alternatively, Pholus may have been only recently expelled from the Kuiper belt, so that surface transformation processes have not yet taken place.

A. Delsanti *et al.* suggest multiple competing processes: reddening by the radiation, and blushing by collisions.[15][16]

## Spectra

The interpretation of spectra is often ambiguous, related to particle sizes and other factors, but the spectra offer an insight into surface composition. As with the colours, the observed spectra can fit a number of models of the surface.

Water ice signatures have been confirmed on a number of centaurs[12] (including 2060 Chiron, 10199 Chariklo and 5145 Pholus). In addition to the water ice signature, a number of other models have been put forward:

- Chariklo's surface has been suggested to be a mixture of tholins (like those detected on Titan and Triton) with amorphous carbon.
- Pholus has been suggested to be covered by a mixture of Titan-like tholins, carbon black, olivine[17] and methanol ice.
- The surface of 52872 Okyrhoe has been suggested to be a mixture of kerogens, olivines and small percentage of water ice.
- 8405 Asbolus has been suggested to be a mixture of 15% Triton-like tholins, 8% Titan-like tholin, 37% amorphous carbon and 40% ice tholin.

Chiron, the only centaur with known cometary activity, appears to be the most complex. The spectra observed vary depending on the period of the observation. Water ice signature was detected during a period of low activity and disappeared during high activity.[19] [20] [21]

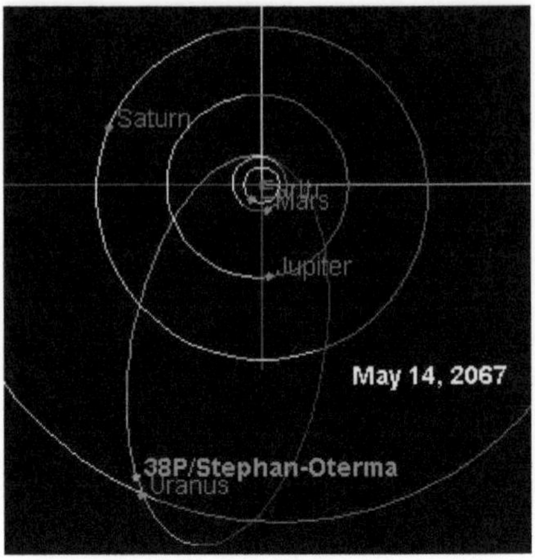

Comet 38P exhibits centaur-like behavior by making close approaches to Jupiter, Saturn, and Uranus between 1982 and 2067.[18]

### Similarities to comets

Observations of Chiron in 1988 and 1989 near its perihelion found it to display a coma (a cloud of gas and dust evaporating from its surface). It is thus now officially classified as both a comet and an asteroid, although it is far larger than a typical comet and there is some lingering controversy. Other centaurs are being monitored for comet-like activity: so far two, 60558 Echeclus, and 166P/NEAT have shown such behavior. 166P/NEAT was discovered while it exhibited a coma, and so is classified as a comet, though its orbit is that of a centaur. 60558 Echeclus was discovered without a coma but recently became active,[22] and so it is now accordingly also classified as both a comet and an asteroid.

There is no clear orbital distinction between centaurs and comets. Both 29P/Schwassmann-Wachmann and 39P/Oterma have been referred to as centaurs since they have typical centaur orbits. The comet 39P/Oterma is currently inactive and was seen to be active only before it was perturbed into a centaur orbit by Jupiter in 1963.[23] The faint comet 38P/Stephan–Oterma would probably not show a coma if it had a perihelion distance beyond Jupiter's orbit at 5 AU. By the year 2200, comet 78P/Gehrels will probably migrate outwards into a centaur-like orbit.

## Theories of origin

The study of centaur development is rich in recent developments but still hampered by limited physical data. Different models have been put forward for possible origin of centaurs.

Simulations indicate that the orbit of some Kuiper-belt objects can be perturbed, resulting in the object's expulsion so that it becomes a centaur. Scattered disk objects would be dynamically the best candidates[24] for such expulsions, but their colours do not fit the bicoloured nature of the centaurs. Plutinos are a class of Kuiper-belt object that display a similar bicoloured nature, and there are suggestions that not all plutinos' orbits are as stable as initially thought, due to perturbation by Pluto.[25] Further developments are expected with more physical data on KBOs.

## Notable centaurs

Well-known centaurs include:

| Name | Year | Discoverer | Half-life[1] (forward) | Class |
|---|---|---|---|---|
| 55576 Amycus | 2002 | NEAT at Palomar | 11.1 Myr | UE |
| 54598 Bienor | 2000 | | | |
| 10370 Hylonome | 1995 | Mauna Kea Observatory | 6.3 Myr | UN |
| 10199 Chariklo | 1997 | Spacewatch | 10.3 Myr | U |
| 8405 Asbolus | 1995 | Spacewatch (James V. Scotti) | 0.86 Myr | SN |
| 7066 Nessus | 1993 | Spacewatch (David L. Rabinowitz) | 4.9 Myr | SE |
| 5145 Pholus | 1992 | Spacewatch (David L. Rabinowitz) | 1.28 Myr | SN |
| 2060 Chiron | 1977 | Charles T. Kowal | 1.03 Myr | SU |

# Notes

[1] Horner, J.; Evans, N.W.; Bailey, M. E. (2004). "Simulations of the Population of Centaurs I: The Bulk Statistics". *Monthly Notices of the Royal Astronomical Society* **354** (3): 798–810. arXiv:astro-ph/0407400. Bibcode 2004MNRAS.354..798H. doi:10.1111/j.1365-2966.2004.08240.x.

[2] "Unusual Minor Planets" (http://www.minorplanetcenter.org/iau/lists/Unusual.html). Minor Planet Center. . Retrieved 2010-10-25.

[3] "Orbit Classification (Centaur)" (http://ssd.jpl.nasa.gov/sbdb_help.cgi?class=CEN). JPL Solar System Dynamics. . Retrieved 2008-10-13.

[4] Elliot, J.L.; Kern, Buie, Trilling; et al. (2005). "The Deep Ecliptic Survey: A Search for Kuiper Belt Objects and Centaurs. II. Dynamical Classification, the Kuiper Belt Plane, and the Core Population" (http://www.iop.org/EJ/abstract/1538-3881/129/2/1117). *The Astronomical Journal* **129** (2): 1117–1162. Bibcode 2005AJ....129.1117E. doi:10.1086/427395. . Retrieved 2008-09-22.

[5] B. Gladman, B. Marsden, C. VanLaerhoven (2008). "Nomenclature in the Outer Solar System". *In **The Solar System Beyond Neptune**, ISBN 978-0-8165-2755-7*.

[6] Chaing, Eugene; Buie, Grundy, Holman; et al. (2007). "A Brief History of Transneptunian Space". *Protostars and Planets V, B. Reipurth, D. Jewitt, and K. Keil (eds.), University of Arizona Press, Tucson*: 895–911. arXiv:astro-ph/0601654. Bibcode 2006astro.ph..1654C.

[7] "JPL Small-Body Database Search Engine" (http://ssd.jpl.nasa.gov/sbdb_query.cgi). JPL Solar System Dynamics. . Retrieved 2010-12-27.

[8] Grundy, Will; Stansberry, J.A.; Noll, K; Stephens, D.C.; Trilling, D.E.; Kern, S.D.; Spencer, J.R.; Cruikshank, D.P. et al. (2007). "The orbit, mass, size, albedo, and density of (65489) Ceto/Phorcys: A tidally-evolved binary Centaur". *Icarus* **191** (1): 286–297. arXiv:0704.1523. Bibcode 2007Icar..191..286G. doi:10.1016/j.icarus.2007.04.004.

[9] Michael E. Brown. "How many dwarf planets are there in the outer solar system? (updates daily)" (http://www.gps.caltech.edu/~mbrown/dps.html). California Institute of Technology. . Retrieved 2012-01-16.

[10] For the purpose of this diagram, an object is classified as a centaur if its semi-major axis lies between Jupiter and Neptune. Last update: October 2008

[11] "Three clones of Centaur 8405 Asbolus making passes within 450Gm" (http://home.surewest.net/kheider/astro/AsbolusClones.txt). . Retrieved 2009-05-02. (Solex 10) (http://chemistry.unina.it/~alvitagl/solex/)

[12] Jewitt, David C.; A. Delsanti (2006). "The Solar System Beyond The Planets". *Solar System Update : Topical and Timely Reviews in Solar System Sciences*. Springer-Praxis Ed.. ISBN 3-540-26056-0. ( Preprint version (pdf) (http://www.ifa.hawaii.edu/faculty/jewitt/papers/2006/DJ06.pdf))

[13] M. A. Barucci, A. Doressoundiram, and D. P. Cruikshank, "Physical Characteristics of TNOs and Centaurs" (2003), available on the web (http://www.lesia.obspm.fr/~alaind/TNO/Barucci2003_comet2.pdf) (accessed 3/20/2008)

[14] Bauer, J. M., Fernández, Y. R., & Meech, K. J. 2003. " An Optical Survey of the Active Centaur C/NEAT (2001 T4) (http://www.journals.uchicago.edu/PASP/journal/issues/v115n810/203101/203101.html)", Publication of the Astronomical Society of the Pacific", **115**, 981

[15] Peixinho, N.; Doressoundiram, A.; Delsanti, A.; Boehnhardt, H.; Barucci, M. A.; Belskaya, I. (2003). "Reopening the TNOs Color Controversy: Centaurs Bimodality and TNOs Unimodality". *Astronomy and Astrophysics* **410** (3): L29–L32. arXiv:astro-ph/0309428. Bibcode 2003A&A...410L..29P. doi:10.1051/0004-6361:20031420.

[16] Hainaut & Delsanti (2002) *Color of Minor Bodies in the Outer Solar System* Astronomy & Astrophysics, **389**, 641 datasource (http://www.sc.eso.org/~ohainaut/MBOSS)

[17] A class of Magnesium Iron Silicates $(Mg, Fe)_2SiO_4$, common components of igneous rocks.

[18] "JPL Close-Approach Data: 38P/Stephan-Oterma" (http://ssd.jpl.nasa.gov/sbdb.cgi?sstr=38P;cad=1#cad). 1981-04-04 last obs. . Retrieved 2009-05-07.

[19] Dotto, E; Barucci, M A; De Bergh, C, *Colours and composition of the centaurs*, Earth, Moon, and Planets, **92**, no. 1–4, pp. 157–167. (June 2003)

[20] Luu, Jane X.; Jewitt, David; Trujillo, C. A. (2000). "Water Ice on 2060 Chiron and its Implications for Centaurs and Kuiper Belt Objects". *The Astrophysical Journal* **531** (2): L151–L154. arXiv:astro-ph/0002094. Bibcode 2000ApJ...531L.151L. doi:10.1086/312536. PMID 10688775.

[21] Fernandez, Y. R.; Jewitt, D. C.; Sheppard, S. S. (2002). "Thermal Properties of Centaurs Asbolus and Chiron". *The Astronomical Journal* **123** (2): 1050–1055. arXiv:astro-ph/0111395. Bibcode 2002AJ....123.1050F. doi:10.1086/338436.

[22] Y-J. Choi, P.R. Weissman, and D. Polishook *(60558) 2000 EC_98*, IAU Circ., **8656** (Jan. 2006), 2.

[23] Mazzotta Epifani, E.; Palumbo; Capria; Cremonese;; et al. (2006). "The dust coma of the active Centaur P/2004 A1 (LONEOS): a CO-driven environment?" (http://google.com/search?q=cache:www.aanda.org/articles/aa/ps/2006/48/aa5189-06.ps.gz). *Astronomy & Astrophysics* **460** (3): 935–944. Bibcode 2006A&A...460..935M. doi:10.1051/0004-6361:20065189. . Retrieved 2009-05-08.

[24] for instance, the centaurs could be part of an "inner" scattered disc of objects perturbed inwards from the Kuiper belt (http://www.zanestein.com/page4_1.htm).

[25] Wan, X.-S; Huang, T.-Y. (2001). "The orbit evolution of 32 plutinos over 100 million year". *Astronomy and Astrophysics* **368** (2): 700–705. Bibcode 2001A&A...368..700W. doi:10.1051/0004-6361:20010056.

## See also

- Asteroid
- Dwarf planet
- Minor planet

## References

## External links

- List of Centaurs and Scattered-Disk Objects (http://www.minorplanetcenter.org/iau/lists/Centaurs.html)
- Centaurs from The Encyclopedia of Astrobiology Astronomy and Spaceflight (http://www.daviddarling.info/encyclopedia/C/Centaur.html)
- Horner, Jonathan; Lykawka, Patryk Sofia (2010). "Planetary Trojans – the main source of short period comets?". *International Journal of Astrobiology* **9** (04): 227–234. arXiv:1007.2541. Bibcode 2010IJAsB...9..227H. doi:10.1017/S1473550410000212.

# Trojan_(astronomy)

In astronomy, a **trojan** is a minor planet or natural satellite (moon) that shares an orbit with a planet or larger moon, but does not collide with it because it orbits around one of the two Lagrangian points of stability (trojan points), $L_4$ and $L_5$, which lie approximately 60° ahead of and behind the larger body, respectively. It is also sometimes called a **Lagrangian** object. Trojan objects are one type of co-orbital object. In this arrangement, the massive star and the smaller planet orbit about their common barycenter—a location in space where the forces of their mutual gravitational attraction balance each other out. A much smaller mass located at one of the Lagrangian points is subject to a combined gravitational force that acts through this barycenter. Hence the object can orbit around the barycenter with the same orbital period as the planet, and the arrangement can remain stable over time.[1]

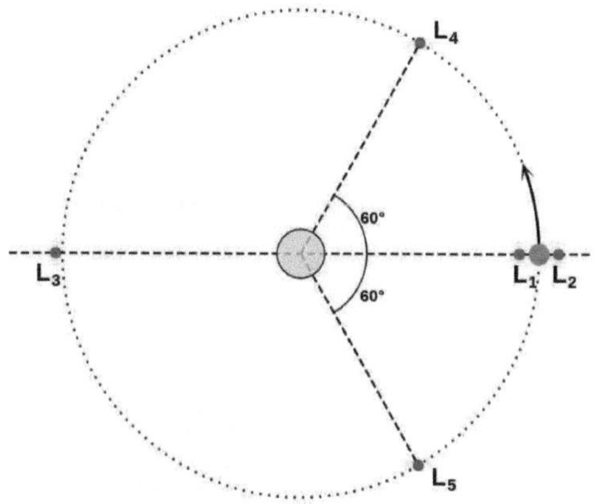


The **trojan points** are those labelled $L_4$ and $L_5$, highlighted in red, on the orbital path of the secondary object (blue), around the primary object (yellow).

Trojan asteroids are asteroids that orbit near one of the two trojan points of a planet. A trojan moon is a moon residing at the trojan point of another (larger) moon. Trojan planets are theoretical planets that orbit at the trojan points of other planets.

Saturn has the only known trojan satellites: Saturn's moon Tethys has two trojan moons (Telesto and Calypso), and Dione also has two trojan moons (Helene and Polydeuces).

In 2011, NASA announced the discovery of the first known Earth trojan.[2]

Numerical calculations of the orbital dynamics involved indicate that Saturn and Uranus probably do not have any primordial trojan asteroids.[3]

## Trojan asteroids

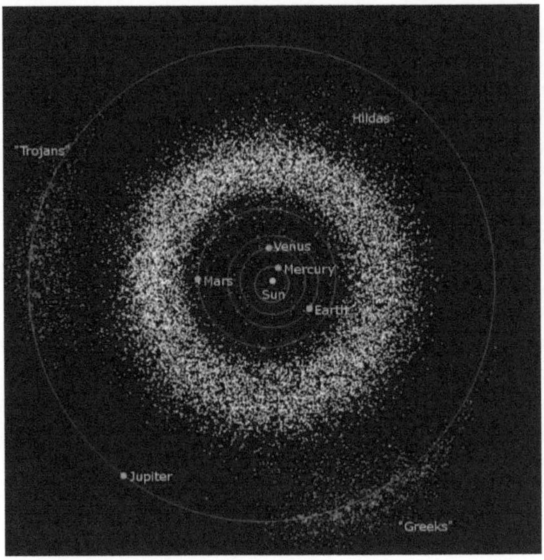

Trojan asteroids of Jupiter (coloured green) in front of and behind the planet along its orbital path. Also shown is the asteroid belt between the orbits of Mars and Jupiter (white), and the Hilda family of asteroids (brown).

In 1772, the Italian–French mathematician and astronomer Joseph-Louis Lagrange obtained two constant-pattern solutions (collinear and equilateral) of the general three-body problem.[4] In the restricted three-body problem, with one mass negligible (which Lagrange did not consider), the five possible positions of that mass are now termed Lagrangian points.

The term "trojan" originally referred to the "trojan asteroids" that orbit close to the Lagrangian points of Jupiter. These have long been named after characters from the Trojan War of Greek mythology. By convention, the asteroids orbiting near the $L_4$ point of Jupiter are named after the characters from the Greek side of the war, whereas those orbiting near the $L_5$ of Jupiter are from the Trojan side. There are two exceptions that were named before the convention was put in place, the Greek-themed 617 Patroclus and the Trojan-themed 624 Hektor, which were assigned to the wrong sides.[5]

Astronomers estimate that the Jupiter trojans are about as numerous as the asteroids of the asteroid belt.[6]

Later on, objects were found orbiting near the Lagrangian points of Neptune, Mars, and Earth.[7] Asteroids at the Lagrangian points of planets other than Jupiter may be called Lagrangian asteroids.[8]

- Three Mars trojans are known: 5261 Eureka, 1998 $VF_{31}$, and 1999 $UJ_7$.[9] 2007 $NS_2$ is possibly one.
- Nine Neptune trojans[10] are known, but they have been estimated to outnumber the Jupiter trojans by an order of magnitude. [11] [12]
- 2010 $TK_7$ was confirmed to be the first known Earth trojan in 2011. It is located in the $L_4$ Lagrangian point, which lies ahead of Earth.[13]

## See also

- Lissajous orbit
- List of objects at Lagrangian points

## References

[1] Robutel, Philippe; Souchay, Jean (2010), "An introduction to the dynamics of trojan asteroids" (http://books.google.com/books?id=CLUYgQlWz4IC&pg=PA197), in Dvorak, Rudolf; Souchay, Jean, *Dynamics of Small Solar System Bodies and Exoplanets*, Lecture Notes in Physics, **790**, Springer, p. 197, ISBN 3-642-04457-3,

[2] NASA's WISE Mission Finds First Trojan Asteroid Sharing Earth's Orbit 7.27.11 (http://www.nasa.gov/mission_pages/WISE/news/wise20110727.html)

[3] Sheppard, Scott S.; Trujillo, Chadwick A. (June 2006). "A Thick Cloud of Neptune Trojans and their Colors" (http://www.dtm.ciw.edu/users/sheppard/pub/Sheppard06NepTroj.pdf) (PDF). *Science* **313** (5786): 511–514. Bibcode 2006Sci...313..511S. doi:10.1126/science.1127173. PMID 16778021. . Retrieved 2008-02-26.

[4] *Essai sur le Problème des Trois Corps* (http://www.ltas-vis.ulg.ac.be/cmsms/uploads/File/Lagrange_essai_3corps.pdf)

[5] Wright, Alison (August 1, 2011). "Planetary science: The Trojan is out there" (http://www.nature.com/nphys/journal/v7/n8/full/nphys2061.html). *Nature Physics* **7** (8): 592. Bibcode 2011NatPh...7..592W. doi:10.1038/nphys2061. . Retrieved 2011-08-12.

[6] Yoshida, Fumi; Nakamura, Tsuko (2005). "Size distribution of faint $L_4$ Trojan asteroids". *The Astronomical journal* **130** (6): 2900–11. Bibcode 2005AJ....130.2900Y. doi:10.1086/497571.

[7] Connors, Martin; Wieger, Paul; Veillet, Christian (27 July 2011). "Earth's Trojan asteroid" (http://www.nature.com/nature/journal/v475/n7357/full/nature10233.html). *Nature* **475** (7357): 481–483. Bibcode 2011Natur.475..481C. doi:10.1038/nature10233. PMID 21796207. . Retrieved 2011-07-27.

[8] Whiteley, Robert J.; and Tholen, David J.; "A CCD Search for Lagrangian Asteroids of the Earth–Sun System", *Icarus* 136:1, November 1998:154–167

[9] "List of Martian Trojans" (http://www.minorplanetcenter.org/iau/lists/MarsTrojans.html). . Retrieved 2012-01-31.

[10] "List of Neptune Trojans" (http://www.minorplanetcenter.org/iau/lists/NeptuneTrojans.html). . Retrieved 2012-12-18.

[11] Chiang, Eugene I. & Lithwick, Yoram; *Neptune Trojans as a Testbed for Planet Formation*, The Astrophysical Journal, **628**, pp. 520–532 Preprint (http://www.arxiv.org/abs/astro-ph/0502276)

[12] Powell, David (30 January 2007). "Neptune May Have Thousands of Escorts" (http://www.space.com/scienceastronomy/070130_st_neptune_trojans.html). Space.com. . Retrieved 2007-03-08.

[13] Choi, Charles Q. (27 July 2011). "First Asteroid Companion of Earth Discovered at Last" (http://www.space.com/12443-earth-asteroid-companion-discovered-2010-tk7.html). Space.com. . Retrieved 2011-07-27.

# Solar_System

Planets and declared dwarf planets of the Solar System. Sizes are to scale. Distances from the Sun are not to scale.

| | |
|---|---|
| **Age** | 4.568 billion years |
| **Location** | Local Interstellar Cloud, Local Bubble, Orion–Cygnus Arm, Milky Way |
| **System mass** | 1.0014 solar masses |
| **Nearest star** | Proxima Centauri (4.22 ly), Alpha Centauri system (4.37 ly) |
| **Nearest known planetary system** | Alpha Centauri system (4.37 ly) |
| **Planetary system** | |
| **Semi-major axis of outer planet (Neptune)** | 4.503 billion km (30.10 AU) |
| **Distance to Kuiper cliff** | 50 AU |
| **No. of stars** | 1<br>Sun |
| **No. of planets** | 8<br>Mercury, Venus, Earth, Mars, Jupiter, Saturn, Uranus, Neptune |
| **No. of known dwarf planets** | 5 (IAU)<br>Ceres, Pluto, Haumea, Makemake, Eris<br>hundreds more candidates[1] |
| **No. of known natural satellites** | 406 (176 of planets[2] and 230 of minor planets[3] ) |
| **No. of known minor planets** | 600,642 (as of 2012-12-13)[2] |
| **No. of known comets** | 3,179 (as of 2012-12-13)[2] |
| **No. of identified round satellites** | 19 |
| **Orbit about the Galactic Center** | |
| **Inclination of invariable plane to the galactic plane** | 60.19° (ecliptic) |
| **Distance to Galactic Center** | 27,000±1,000 ly |
| **Orbital speed** | 220 km/s |
| **Orbital period** | 225–250 Myr |
| **Star-related properties** | |
| **Spectral type** | G2V |
| **Frost line** | ≈5 AU[4] |
| **Distance to heliopause** | ≈120 AU |

| Hill sphere radius | ≈1–2 ly |
|---|---|

The **Solar System**[a] consists of the Sun and its planetary system of eight planets, their moons, and other non-stellar objects.[b] It formed 4.6 billion years ago from the collapse of a giant molecular cloud. The vast majority of the system's mass is in the Sun, with most of the remaining mass contained in Jupiter. The four smaller inner planets, Mercury, Venus, Earth and Mars, also called the terrestrial planets, are primarily composed of rock and metal. The four outer planets, called the gas giants, are substantially more massive than the terrestrials. The two largest, Jupiter and Saturn, are composed mainly of hydrogen and helium; the two outermost planets, Uranus and Neptune, are composed largely of substances with relatively high melting points (compared with hydrogen and helium), called *ices*, such as water, ammonia and methane, and are often referred to separately as "ice giants". All planets have almost circular orbits that lie within a nearly flat disc called the ecliptic plane.

The Solar System also contains a number of regions populated by smaller objects.[b] The asteroid belt, which lies between Mars and Jupiter, is similar to the terrestrial planets as it mostly contains objects composed of rock and metal. Beyond Neptune's orbit lie the Kuiper belt and scattered disc; linked populations of trans-Neptunian objects composed mostly of ices. Within these populations, several dozen to more than ten thousand objects may be large enough to have been rounded by their own gravity.[5] Such objects are referred to as dwarf planets. Identified dwarf planets include the asteroid Ceres and the trans-Neptunian objects Pluto, Eris, Haumea, and Makemake.[b] In addition to these two regions, various other small-body populations including comets, centaurs and interplanetary dust freely travel between regions. Six of the planets, at least three of the dwarf planets, and many of the smaller bodies are orbited by natural satellites,[c] usually termed "moons" after Earth's Moon. Each of the outer planets is encircled by planetary rings of dust and other small objects.

The solar wind, a flow of plasma from the Sun, creates a bubble in the interstellar medium known as the heliosphere, which extends out to the edge of the scattered disc. The Oort cloud, which is believed to be the source for long-period comets, may also exist at a distance roughly a thousand times further than the heliosphere. The heliopause is the point at which pressure from the solar wind is equal to the opposing pressure of interstellar wind. The Solar System is located within one of the outer arms of Milky Way galaxy, which contains about 200 billion stars.

## Discovery and exploration

For many thousands of years, humanity, with a few notable exceptions, did not recognize the existence of the Solar System. People believed the Earth to be stationary at the centre of the universe and categorically different from the divine or ethereal objects that moved through the sky. Although the Greek philosopher Aristarchus of Samos had speculated on a heliocentric reordering of the cosmos,[6] Nicolaus Copernicus was the first to develop a mathematically predictive heliocentric system.[7] His 17th-century successors, Galileo Galilei, Johannes Kepler and Isaac Newton, developed an understanding of physics that led to the gradual acceptance of the idea that the Earth moves around the Sun and that the planets are governed by the same physical laws that governed the Earth. Additionally, the invention of the telescope led to the discovery of further planets and moons. In more recent times, improvements in the telescope and the use of unmanned spacecraft have enabled the investigation of geological phenomena such as mountains and craters, and seasonal meteorological phenomena such as clouds, dust storms and ice caps on the other planets.

## Structure and composition

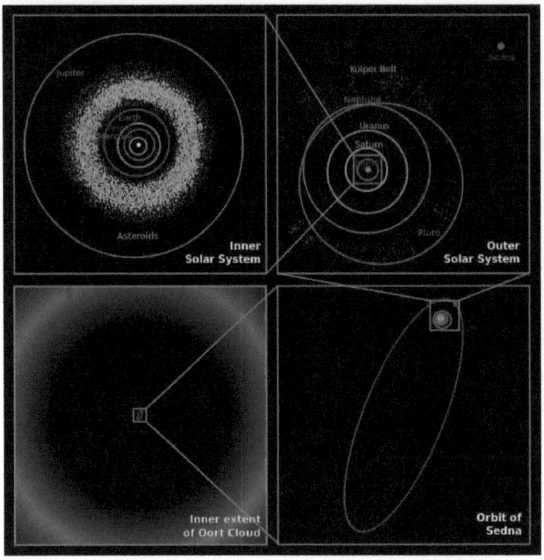

The orbits of the bodies in the Solar System to scale (clockwise from top left)

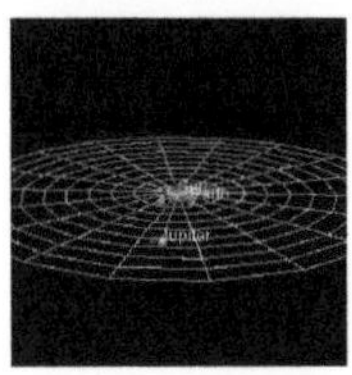

The principal component of the Solar System is the Sun, a G2 main-sequence star that contains 99.86 percent of the system's known mass and dominates it gravitationally.[8] The Sun's four largest orbiting bodies, the gas giants, account for 99 percent of the remaining mass, with Jupiter and Saturn together comprising more than 90 percent.[d]

Most large objects in orbit around the Sun lie near the plane of Earth's orbit, known as the ecliptic. The planets are very close to the ecliptic while comets and Kuiper belt objects are frequently at significantly greater angles to it.[9][10] All the planets and most other objects orbit the Sun in the same direction that the Sun is rotating (counter-clockwise, as viewed from above the Sun's north pole).[11] There are exceptions, such as Halley's Comet.

The overall structure of the charted regions of the Solar System consists of the Sun, four relatively small inner planets surrounded by a belt of rocky asteroids, and four gas giants surrounded by the Kuiper belt of icy objects. Astronomers sometimes informally divide this structure into separate regions. The *inner Solar System* includes the four terrestrial planets and the asteroid belt. The *outer Solar System* is beyond the asteroids, including the four gas giants.[12] Since the discovery of the Kuiper belt, the outermost parts of the Solar System are considered a distinct region consisting of the objects beyond Neptune.[13]

Most of the planets in the Solar System possess secondary systems of their own, being orbited by planetary objects called natural satellites, or moons (two of which are larger than the planet Mercury), or, in the case of the four gas giants, by planetary rings; thin bands of tiny particles that orbit them in unison. Most of the largest natural satellites are in synchronous rotation, with one face permanently turned toward their parent.

Kepler's laws of planetary motion describe the orbits of objects about the Sun. Following Kepler's laws, each object travels along an ellipse with the Sun at one focus. Objects closer to the Sun (with smaller semi-major axes) travel more quickly, as they are more affected by the Sun's gravity. On an elliptical orbit, a body's distance from the Sun varies over the course of its year. A body's closest approach to the Sun is called its *perihelion*, while its most distant point from the Sun is called its *aphelion*. The orbits of the planets are nearly circular, but many comets, asteroids and Kuiper belt objects follow highly elliptical orbits. The positions of the bodies in the Solar System can be predicted using numerical models.

Although the Sun dominates the system by mass, it accounts for only about 2% of the angular momentum[14] due to the differential rotation within the gaseous Sun.[15] The planets, dominated by Jupiter, account for most of the rest of the angular momentum due to the combination of their mass, orbit, and distance from the Sun, with a possibly significant contribution from comets.[14]

Due to the vast distances involved, many representations of the Solar System show orbits the same distance apart. In reality, with a few exceptions, the farther a planet or belt is from the Sun, the larger the distance between it and the previous orbit. For example, Venus is approximately 0.33 astronomical units (AU)[e] farther out from the Sun than Mercury, while Saturn is 4.3 AU out from Jupiter, and Neptune lies 10.5 AU out from Uranus. Attempts have been made to determine a relationship between these orbital distances (for example, the Titius–Bode law),[16] but no such theory has been accepted.

The Sun, which comprises nearly all the matter in the Solar System, is composed of roughly 98% hydrogen and helium.[17] Jupiter and Saturn, which comprise nearly all the remaining matter, possess atmospheres composed of roughly 99% of those same elements.[18] [19] A composition gradient exists in the Solar System, created by heat and light pressure from the Sun; those objects closer to the Sun, which are more affected by heat and light pressure, are composed of elements with high melting points. Objects farther from the Sun are composed largely of materials with lower melting points.[20] The boundary in the Solar System beyond which those volatile substances could condense is known as the frost line, and it lies at roughly 5 AU from the Sun.[4]

The objects of the inner Solar System are composed mostly of *rock*,[21] the collective name for compounds with high melting points, such as silicates, iron or nickel, that remained solid under almost all conditions in the protoplanetary nebula.[22] Jupiter and Saturn are composed mainly of *gases*, the astronomical term for materials with extremely low melting points and high vapor pressure such as molecular hydrogen, helium, and neon, which were always in the gaseous phase in the nebula.[22] *Ices*, like water, methane, ammonia, hydrogen sulfide and carbon dioxide,[21] have melting points up to a few hundred kelvins, while their phase depends on the ambient pressure and temperature.[22] They can be found as ices, liquids, or gases in various places in the Solar System, while in the nebula they were either in the solid or gaseous phase.[22] Icy substances comprise the majority of the satellites of the giant planets, as well as most of Uranus and Neptune (the so-called "ice giants") and the numerous small objects that lie beyond Neptune's orbit.[21] [23] Together, gases and ices are referred to as *volatiles*.[24]

A number of Solar System models on Earth attempt to convey the relative scales involved in the Solar System on human terms. Some models are mechanical — called orreries — while others extend across cities or regional areas.[25] The largest such scale model, the Sweden Solar System, uses the 110-metre Ericsson Globe in Stockholm as its substitute Sun, and, following the scale, Jupiter is a 7.5 metre sphere at Arlanda International Airport, 40 km away, while the farthest current object, Sedna, is a 10-cm sphere in Luleå, 912 km away.[26] [27]

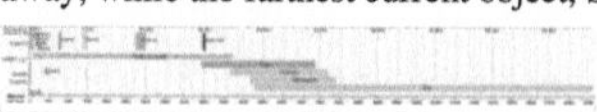

Range of selected bodies of the Solar System from the middle of the Sun. The left and right edges of each bar correspond to the perihelion and aphelion of the body, respectively. Long bars denote high orbital eccentricity.

## Formation and evolution

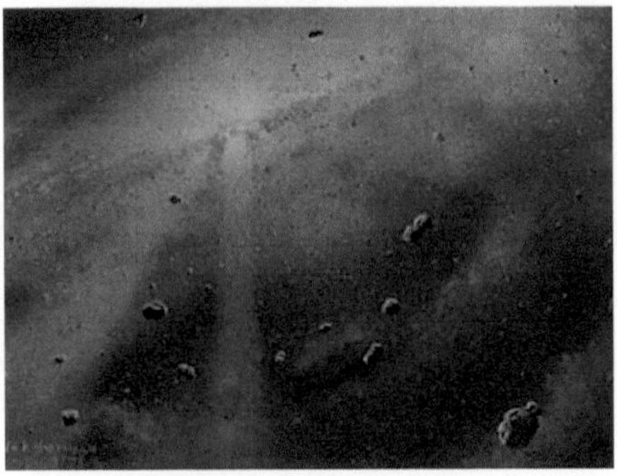
Artist's conception of the early Solar System

The Solar System formed 4.568 billion years ago from the gravitational collapse of a region within a large molecular cloud.[28] This initial cloud was likely several light-years across and probably birthed several stars.[29] As is typical of molecular clouds, this one consisted mostly of hydrogen, with some helium, and small amounts of heavier elements fused by previous generations of stars. As the region that would become the Solar System, known as the pre-solar nebula,[30] collapsed, conservation of angular momentum caused it to rotate faster. The centre, where most of the mass collected, became increasingly hotter than the surrounding disc.[29] As the contracting nebula rotated faster, it began to flatten into a protoplanetary disc with a diameter of roughly 200 AU[29] and a hot, dense protostar at the centre.[31][32] The planets formed by accretion from this disc,[33] in which dust and gas gravitationally attracted each other, coalescing to form ever larger bodies. Hundreds of protoplanets may have existed in the early Solar System, but they either merged or were destroyed, leaving the planets, dwarf planets, and leftover minor bodies.

Due to their higher boiling points, only metals and silicates could exist in the warm inner Solar System close to the Sun, and these would form the rocky planets of Mercury, Venus, Earth, and Mars. Since metallic elements only comprised a very small fraction of the solar nebula, the terrestrial planets could not grow very large. The gas giants (Jupiter, Saturn, Uranus, and Neptune) formed further out, beyond the frost line, the point between the orbits of Mars and Jupiter where material is cool enough for volatile icy compounds to remain solid. The ices that formed these planets were more plentiful than the metals and silicates that formed the terrestrial inner planets, allowing them to grow massive enough to capture large atmospheres of hydrogen and helium, the lightest and most abundant elements. Leftover debris that never became planets congregated in regions such as the asteroid belt, Kuiper belt, and Oort cloud. The Nice model is an explanation for the creation of these regions, and how the outer planets could have formed in different positions and migrated to their current orbits through various gravitational interactions.

Within 50 million years, the pressure and density of hydrogen in the centre of the protostar became great enough for it to begin thermonuclear fusion.[34] The temperature, reaction rate, pressure, and density increased until hydrostatic equilibrium was achieved: the thermal pressure equaled the force of gravity. At this point the Sun became a main-sequence star.[35] Solar wind from the Sun created the heliosphere and swept away the remaining gas and dust from the protoplanetary disc into interstellar space, ending the planetary formation process.

The Solar System will remain roughly as we know it today until the hydrogen in the core of the Sun has been entirely converted to helium, which will occur roughly 5.4 billion years from now. This will mark the end of the Sun's main-sequence life. At this time, the core of the Sun will collapse, and the energy output will be much greater than at present. The outer layers of the Sun will expand to roughly up to 260 times its current diameter and the Sun will become a red giant. Because of its vastly increased surface area, the surface of the Sun will be considerably cooler than it is on the main sequence (2600 K at the coolest).[36] The expanding Sun is expected to vaporize Mercury and Venus and render the Earth uninhabitable, as the habitable zone moves out to the orbit of Mars. Eventually, the core will be hot enough for helium fusion to begin in the core; the Sun will burn helium for a fraction of the time it burned hydrogen in the core. The Sun is not massive enough to commence fusion of heavier elements, and nuclear reactions in the core will dwindle. Its outer layers will fall away into space, leaving a white dwarf, an extraordinarily dense object, half the original mass of the Sun but only the size of the Earth.[37] The ejected outer layers will form what is known as a planetary nebula, returning some of the material that formed the Sun—but now enriched with heavier elements like carbon—to the interstellar medium.

## Sun

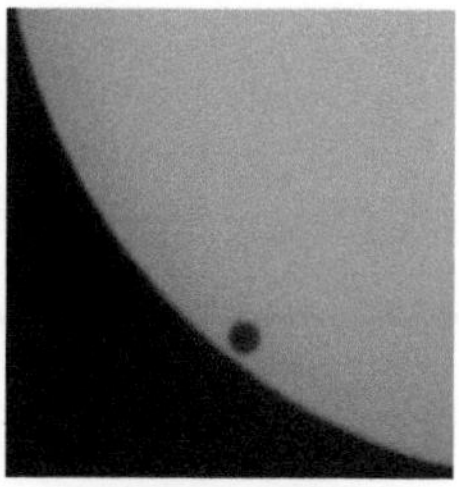

A transit of Venus

The Sun is the Solar System's star, and by far its chief component. Its large mass (332,900 Earth masses)[38] produces temperatures and densities in its core high enough to sustain nuclear fusion,[39] which releases enormous amounts of energy, mostly radiated into space as electromagnetic radiation, peaking in the 400–700 nm band of visible light.[40]

The Sun is classified as a type G2 yellow dwarf, but this name is misleading as, compared to the majority of stars in our galaxy, the Sun is rather large and bright.[41] Stars are classified by the Hertzsprung–Russell diagram, a graph that plots the brightness of stars with their surface temperatures. Generally, hotter stars are brighter. Stars following this pattern are said to be on the main sequence, and the Sun lies right in the middle of it. However, stars brighter and hotter than the Sun are rare, while substantially dimmer and cooler stars, known as red dwarfs, are common, making up 85 percent of the stars in the galaxy.[41] [42]

Evidence suggests that the Sun's position on the main sequence puts it in the "prime of life" for a star, in that it has not yet exhausted its store of hydrogen for nuclear fusion. The Sun is growing brighter; early in its history it was 70 percent as bright as it is today.[43]

The Sun is a population I star; it was born in the later stages of the universe's evolution, and thus contains more elements heavier than hydrogen and helium ("metals" in astronomical parlance) than older population II stars.[44] Elements heavier than hydrogen and helium were formed in the cores of ancient and exploding stars, so the first generation of stars had to die before the universe could be enriched with these atoms. The oldest stars contain few metals, while stars born later have more. This high metallicity is thought to have been crucial to the Sun's developing a planetary system, because planets form from accretion of "metals".[45]

The heliospheric current sheet

## Interplanetary medium

Along with light, the Sun radiates a continuous stream of charged particles (a plasma) known as the solar wind. This stream of particles spreads outwards at roughly 1.5 million kilometres per hour,[46] creating a tenuous atmosphere (the heliosphere) that permeates the Solar System out to at least 100 AU (see heliopause).[47] This is known as the interplanetary medium. Activity on the Sun's surface, such as solar flares and coronal mass ejections, disturb the heliosphere, creating space weather and causing geomagnetic storms.[48] The largest structure within the heliosphere is the heliospheric current sheet, a spiral form created by the actions of the Sun's rotating magnetic field on the interplanetary medium.[49] [50]

Earth's magnetic field stops its atmosphere from being stripped away by the solar wind. Venus and Mars do not have magnetic fields, and as a result, the solar wind causes their atmospheres to gradually bleed away into space.[51] Coronal mass ejections and similar events blow a magnetic field and huge quantities of material from the surface of the Sun. The interaction of this magnetic field and material with Earth's magnetic field funnels charged particles into the Earth's upper atmosphere, where its interactions create aurorae seen near the magnetic poles.

Cosmic rays originate outside the Solar System. The heliosphere partially shields the Solar System, and planetary magnetic fields (for those planets that have them) also provide some protection. The density of cosmic rays in the interstellar medium and the strength of the Sun's magnetic field change on very long timescales, so the level of cosmic radiation in the Solar System varies, though by how much is unknown.[52]

The interplanetary medium is home to at least two disc-like regions of cosmic dust. The first, the zodiacal dust cloud, lies in the inner Solar System and causes zodiacal light. It was likely formed by collisions within the asteroid belt brought on by interactions with the planets.[53] The second extends from about 10 AU to about 40 AU, and was probably created by similar collisions within the Kuiper belt.[54] [55]

# Inner Solar System

The inner Solar System is the traditional name for the region comprising the terrestrial planets and asteroids.[56] Composed mainly of silicates and metals, the objects of the inner Solar System are relatively close to the Sun; the radius of this entire region is shorter than the distance between Jupiter and Saturn.

## Inner planets

The four inner or terrestrial planets have dense, rocky compositions, few or no moons, and no ring systems. They are composed largely of refractory minerals, such as the silicates, which form their crusts and mantles, and metals such as iron and nickel, which form their cores. Three of the four inner planets (Venus, Earth and Mars) have atmospheres substantial enough to generate weather; all have impact craters and tectonic surface features such as rift valleys and volcanoes. The term *inner planet* should not be confused with *inferior planet*, which designates those planets that are closer to the Sun than Earth is (i.e. Mercury and Venus).

The inner planets. From left to right: Mercury, Venus, Earth, and Mars in true-colour (sizes to scale, interplanetary distances not)

### Mercury

Mercury (0.4 AU from the Sun) is the closest planet to the Sun and the smallest planet in the Solar System (0.055 Earth masses). Mercury has no natural satellites, and its only known geological features besides impact craters are lobed ridges or rupes, probably produced by a period of contraction early in its history.[57] Mercury's almost negligible atmosphere consists of atoms blasted off its surface by the solar wind.[58] Its relatively large iron core and thin mantle have not yet been adequately explained. Hypotheses include that its outer layers were stripped off by a giant impact, and that it was prevented from fully accreting by the young Sun's energy.[59] [60]

### Venus

Venus (0.7 AU from the Sun) is close in size to Earth (0.815 Earth masses), and, like Earth, has a thick silicate mantle around an iron core, a substantial atmosphere and evidence of internal geological activity. However, it is much drier than Earth and its atmosphere is ninety times as dense. Venus has no natural satellites. It is the hottest planet, with surface temperatures over 400 °C, most likely due to the amount of greenhouse gases in the atmosphere.[61] No definitive evidence of current geological activity has been detected on Venus, but it has no magnetic field that would prevent depletion of its substantial atmosphere, which suggests that its atmosphere is regularly replenished by volcanic eruptions.[62]

### Earth

Earth (1 AU from the Sun) is the largest and densest of the inner planets, the only one known to have current geological activity, and is the only place in the Solar System where life is known to exist.[63] Its liquid hydrosphere is unique among the terrestrial planets, and it is also the only planet where plate tectonics has been observed. Earth's atmosphere is radically different from those of the other planets, having been altered by the presence of life to contain 21% free oxygen.[64] It has one natural satellite, the Moon, the only large satellite of a terrestrial planet in the Solar System.

### Mars

Mars (1.5 AU from the Sun) is smaller than Earth and Venus (0.107 Earth masses). It possesses an atmosphere of mostly carbon dioxide with a surface pressure of 6.1 millibars (roughly 0.6 percent that of the Earth's).[65] Its surface, peppered with vast volcanoes such as Olympus Mons and rift valleys such as Valles Marineris, shows geological activity that may have persisted until as recently as 2 million years ago.[66] Its red colour comes from iron oxide (rust) in its soil.[67] Mars has two tiny natural satellites (Deimos and Phobos) thought to be captured asteroids.[68]

## Asteroid belt

Asteroids are small Solar System bodies[b] composed mainly of refractory rocky and metallic minerals, with some ice.[69]

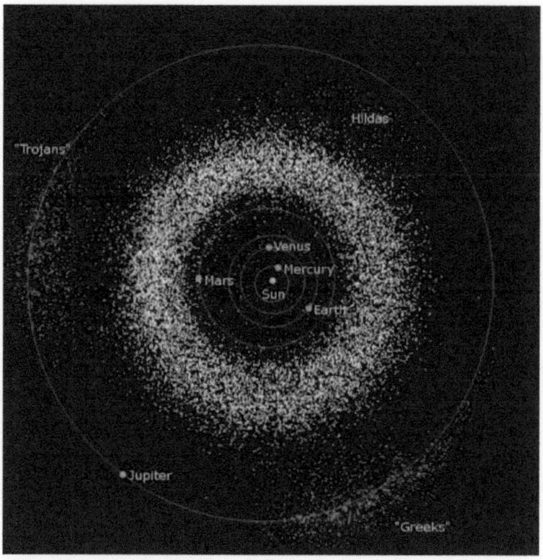

Image of the asteroid belt (white), the Jupiter Trojans (green), the Hildas (orange), and near-Earth asteroids.

The asteroid belt occupies the orbit between Mars and Jupiter, between 2.3 and 3.3 AU from the Sun. It is thought to be remnants from the Solar System's formation that failed to coalesce because of the gravitational interference of Jupiter.[70]

Asteroids range in size from hundreds of kilometres across to microscopic. All asteroids except the largest, Ceres, are classified as small Solar System bodies.[71]

The asteroid belt contains tens of thousands, possibly millions, of objects over one kilometre in diameter.[72] Despite this, the total mass of the asteroid belt is unlikely to be more than a thousandth of that of the Earth.[73] The asteroid belt is very sparsely populated; spacecraft routinely pass through without incident. Asteroids with diameters between 10 and $10^{-4}$ m are called meteoroids.[74]

### Ceres

Ceres (2.77 AU) is the largest asteroid, a protoplanet, and a dwarf planet.[b] It has a diameter of slightly under 1000 km, and a mass large enough for its own gravity to pull it into a spherical shape. Ceres was considered a planet when it was discovered in the 19th century, but was reclassified as an asteroid in the 1850s as further observations revealed additional asteroids.[75] It was classified as a dwarf planet in 2006.

### Asteroid groups

Asteroids in the asteroid belt are divided into asteroid groups and families based on their orbital characteristics. Asteroid moons are asteroids that orbit larger asteroids. They are not as clearly distinguished as planetary moons, sometimes being almost as large as their partners. The asteroid belt also contains main-belt comets, which may have been the source of Earth's water.[76]

Jupiter trojans are located in either of Jupiter's $L_4$ or $L_5$ points (gravitationally stable regions leading and trailing a planet in its orbit); the term "Trojan" is also used for small bodies in any other planetary or satellite Lagrange point. Hilda asteroids are in a 2:3 resonance with Jupiter; that is, they go around the Sun three times for every two Jupiter orbits.[77]

The inner Solar System is also dusted with rogue asteroids, many of which cross the orbits of the inner planets.[78]

# Outer Solar System

The outer region of the Solar System is home to the gas giants and their large moons. Many short-period comets, including the centaurs, also orbit in this region. Due to their greater distance from the Sun, the solid objects in the outer Solar System contain a higher proportion of volatiles such as water, ammonia and methane, than the rocky denizens of the inner Solar System, as the colder temperatures allow these compounds to remain solid.

## Outer planets

The four outer planets, or gas giants (sometimes called Jovian planets), collectively make up 99 percent of the mass known to orbit the Sun.[d] Jupiter and Saturn are each many tens of times the mass of the Earth and consist overwhelmingly of hydrogen and helium; Uranus and Neptune are far less massive (<20 Earth masses) and possess more ices in their makeup. For these reasons, some astronomers suggest they belong in their own category, "ice giants".[79] All four gas giants have rings, although only Saturn's ring system is easily observed from Earth. The term *outer planet* should not be confused with *superior planet*, which designates planets outside Earth's orbit and thus includes both the outer planets and Mars.

From top to bottom: Neptune, Uranus, Saturn, and Jupiter (not to scale)

### Jupiter

Jupiter (5.2 AU), at 318 Earth masses, is 2.5 times the mass of all the other planets put together. It is composed largely of hydrogen and helium. Jupiter's strong internal heat creates a number of semi-permanent features in its atmosphere, such as cloud bands and the Great Red Spot.

Jupiter has 67 known satellites. The four largest, Ganymede, Callisto, Io, and Europa, show similarities to the terrestrial planets, such as volcanism and internal heating.[80] Ganymede, the largest satellite in the Solar System, is larger than Mercury.

### Saturn

Saturn (9.5 AU), distinguished by its extensive ring system, has several similarities to Jupiter, such as its atmospheric composition and magnetosphere. Although Saturn has 60% of Jupiter's volume, it is less than a third as massive, at 95 Earth masses, making it the least dense planet in the Solar System. The rings of Saturn are made up of small ice and rock particles.

Saturn has 62 confirmed satellites; two of which, Titan and Enceladus, show signs of geological activity, though they are largely made of ice.[81] Titan, the second-largest moon in the Solar System, is larger than Mercury and the only known satellite in the Solar System with a substantial atmosphere.

### Uranus

Uranus (19.6 AU), at 14 Earth masses, is the lightest of the outer planets. Uniquely among the planets, it orbits the Sun on its side; its axial tilt is over ninety degrees to the ecliptic. It has a much colder core than the other gas giants, and radiates very little heat into space.[82]

Uranus has 27 known satellites, the largest ones being Titania, Oberon, Umbriel, Ariel and Miranda.

### Neptune

Neptune (30 AU), though slightly smaller than Uranus, is more massive (equivalent to 17 Earths) and therefore more dense. It radiates more internal heat, but not as much as Jupiter or Saturn.[83]

Neptune has 13 known satellites. The largest, Triton, is geologically active, with geysers of liquid nitrogen.[84] Triton is the only large satellite with a retrograde orbit. Neptune is accompanied in its orbit by a number of minor planets, termed Neptune trojans, that are in 1:1 resonance with it.

## Centaurs

The centaurs are icy comet-like bodies with a semi-major axis greater than Jupiter's (5.5 AU) and less than Neptune's (30 AU). The largest known centaur, 10199 Chariklo, has a diameter of about 250 km.[85] The first centaur discovered, 2060 Chiron, has also been classified as comet (95P) since it develops a coma just as comets do when they approach the Sun.[86]

# Comets

Comet Hale–Bopp

Comets are small Solar System bodies,[b] typically only a few kilometres across, composed largely of volatile ices. They have highly eccentric orbits, generally a perihelion within the orbits of the inner planets and an aphelion far beyond Pluto. When a comet enters the inner Solar System, its proximity to the Sun causes its icy surface to sublimate and ionise, creating a coma: a long tail of gas and dust often visible to the naked eye.

Short-period comets have orbits lasting less than two hundred years. Long-period comets have orbits lasting thousands of years. Short-period comets are believed to originate in the Kuiper belt, while long-period comets, such as Hale–Bopp, are believed to originate in the Oort cloud. Many comet groups, such as the Kreutz Sungrazers, formed from the breakup of a single parent.[87] Some comets with hyperbolic orbits may originate outside the Solar System, but determining their precise orbits is difficult.[88] Old comets that have had most of their volatiles driven out by solar warming are often categorised as asteroids.[89]

## Trans-Neptunian region

The area beyond Neptune, or the "trans-Neptunian region", is still largely unexplored. It appears to consist overwhelmingly of small worlds (the largest having a diameter only a fifth that of the Earth and a mass far smaller than that of the Moon) composed mainly of rock and ice. This region is sometimes known as the "outer Solar System", though others use that term to mean the region beyond the asteroid belt.

### Kuiper belt

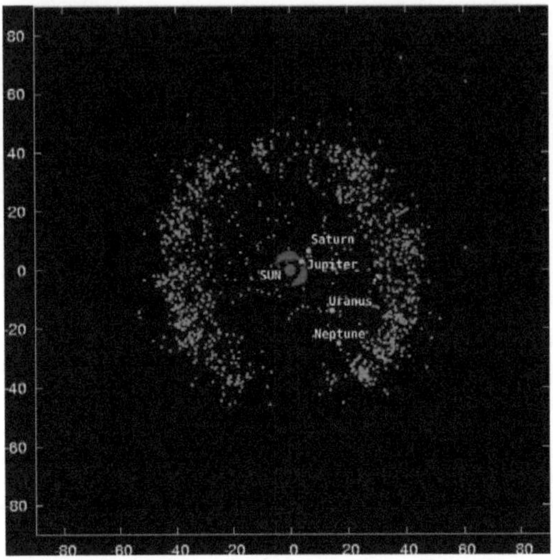

Plot of all known Kuiper belt objects, set against the four outer planets

The Kuiper belt, the region's first formation, is a great ring of debris similar to the asteroid belt, but composed mainly of ice.[90] It extends between 30 and 50 AU from the Sun. Though it is thought to contain dozens of dwarf planets, it is composed mainly of small Solar System bodies. Many of the larger Kuiper belt objects, such as Quaoar, Varuna, and Orcus, may be recognized as dwarf planets with further data. There are estimated to be over 100,000 Kuiper belt objects with a diameter greater than 50 km, but the total mass of the Kuiper belt is thought to be only a tenth or even a hundredth the mass of the Earth.[91] Many Kuiper belt objects have multiple satellites,[92] and most have orbits that take them outside the plane of the ecliptic.[93]

The Kuiper belt can be roughly divided into the "classical" belt and the resonances.[90] Resonances are orbits linked to that of Neptune (e.g. twice for every three Neptune orbits, or once for every two). The first resonance begins within the orbit of Neptune itself. The classical belt consists of objects having no resonance with Neptune, and extends from roughly 39.4 AU to 47.7 AU.[94] Members of the classical Kuiper belt are classified as cubewanos, after the first of their kind to be discovered, (15760) 1992 $QB_1$, and are still in near primordial, low-eccentricity orbits.[95]

### Pluto and Charon

The dwarf planet Pluto (39 AU average) is the largest known object in the Kuiper belt. When discovered in 1930, it was considered to be the ninth planet; this changed in 2006 with the adoption of a formal definition of planet. Pluto has a relatively eccentric orbit inclined 17 degrees to the ecliptic plane and ranging from 29.7 AU from the Sun at perihelion (within the orbit of Neptune) to 49.5 AU at aphelion.

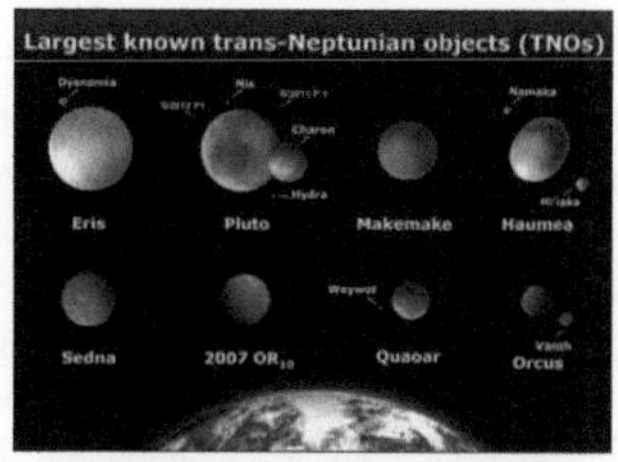


10}}, Quaoar, Orcus, and Earth. These eight trans-Neptunian objects have the brightest absolute magnitudes, although several other TNOs have been found to be physically larger than Orcus, and several more may yet be found.

Charon, Pluto's largest moon, is sometimes described as part of a binary system with Pluto, as the two bodies orbit a barycentre of gravity above their surfaces (i.e., they appear to "orbit each other"). Beyond Charon, four much smaller moons, P5, Nix, P4, and Hydra are known to orbit within the system.

Pluto has a 3:2 resonance with Neptune, meaning that Pluto orbits twice round the Sun for every three Neptunian orbits. Kuiper belt objects whose orbits share this resonance are called plutinos.[96]

### Makemake and Haumea

Makemake (45.79 AU average), while smaller than Pluto, is the largest known object in the *classical* Kuiper belt (that is, it is not in a confirmed resonance with Neptune). Makemake is the brightest object in the Kuiper belt after Pluto. It was named and designated a dwarf planet in 2008.[97] Its orbit is far more inclined than Pluto's, at 29°.[98]

Haumea (43.13 AU average) is in an orbit similar to Makemake except that it is caught in a 7:12 orbital resonance with Neptune.[99] It is about the same size as Makemake and has two natural satellites. A rapid, 3.9-hour rotation gives it a flattened and elongated shape. It was named and designated a dwarf planet in 2008. [100]

## Scattered disc

The scattered disc, which overlaps the Kuiper belt but extends much further outwards, is thought to be the source of short-period comets. Scattered disc objects are believed to have been ejected into erratic orbits by the gravitational influence of Neptune's early outward migration. Most scattered disc objects (SDOs) have perihelia within the Kuiper belt but aphelia far beyond it (some have aphelia farther than 150 AU from the Sun). SDOs' orbits are also highly inclined to the ecliptic plane, and are often almost perpendicular to it. Some astronomers consider the scattered disc to be merely another region of the Kuiper belt, and describe scattered disc objects as "scattered Kuiper belt objects".[101] Some astronomers also classify centaurs as inward-scattered Kuiper belt objects along with the outward-scattered residents of the scattered disc.[102]

### Eris

Eris (68 AU average) is the largest known scattered disc object, and caused a debate about what constitutes a planet, since it is 25% more massive than Pluto[103] and about the same diameter. It is the most massive of the known dwarf planets. It has one known moon, Dysnomia. Like Pluto, its orbit is highly eccentric, with a perihelion of 38.2 AU (roughly Pluto's distance from the Sun) and an aphelion of 97.6 AU, and steeply inclined to the ecliptic plane.

# Farthest regions

The point at which the Solar System ends and interstellar space begins is not precisely defined, since its outer boundaries are shaped by two separate forces: the solar wind and the Sun's gravity. The outer limit of the solar wind's influence is roughly four times Pluto's distance from the Sun; this *heliopause* is considered the beginning of the interstellar medium.[47] However, the Sun's Hill sphere, the effective range of its gravitational dominance, is believed to extend up to a thousand times farther.[104]

## Heliopause

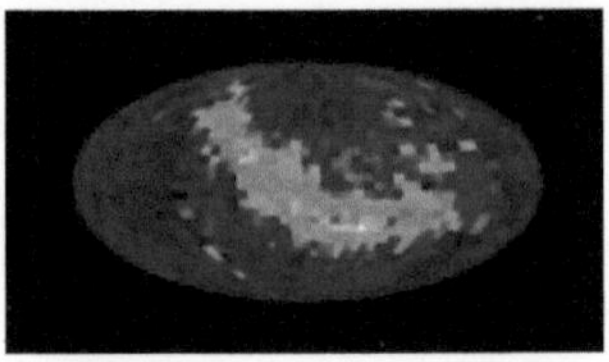

Energetic neutral atoms map of heliosheath and heliopause by IBEX. Credit: NASA/Goddard Space Flight Center Scientific Visualization Studio.

The heliosphere is divided into two separate regions. The solar wind travels at roughly 400 km/s until it collides with the interstellar wind; the flow of plasma in the interstellar medium. The collision occurs at the termination shock, which is roughly 80–100 AU from the Sun upwind of the interstellar medium and roughly 200 AU from the Sun downwind.[105] Here the wind slows dramatically, condenses and becomes more turbulent,[105] forming a great oval structure known as the heliosheath. This structure is believed to look and behave very much like a comet's tail, extending outward for a further 40 AU on the upwind side but tailing many times that distance downwind; but evidence from the Cassini and Interstellar Boundary Explorer spacecraft has suggested that it is in fact forced into a bubble shape by the constraining action of the interstellar magnetic field.[106] Both *Voyager 1* and *Voyager 2* are reported to have passed the termination shock and entered the heliosheath, at 94 and 84 AU from the Sun, respectively.[107] [108] The outer boundary of the heliosphere, the heliopause, is the point at which the solar wind finally terminates and is the beginning of interstellar space.[47]

The shape and form of the outer edge of the heliosphere is likely affected by the fluid dynamics of interactions with the interstellar medium[105] as well as solar magnetic fields prevailing to the south, e.g. it is bluntly shaped with the northern hemisphere extending 9 AU farther than the southern hemisphere. Beyond the heliopause, at around 230 AU, lies the bow shock, a plasma "wake" left by the Sun as it travels through the Milky Way.[109]

No spacecraft have yet passed beyond the heliopause, so it is impossible to know for certain the conditions in local interstellar space. It is expected that NASA's Voyager spacecraft will pass the heliopause some time in the next decade and transmit valuable data on radiation levels and solar wind back to the Earth.[110] How well the heliosphere shields the Solar System from cosmic rays is poorly understood. A NASA-funded team has developed a concept of a "Vision Mission" dedicated to sending a probe to the heliosphere.[111] [112]

## Sedna

90377 Sedna (525.86 AU average) is a large, reddish object with a gigantic, highly elliptical orbit that takes it from about 76 AU at perihelion to 928 AU at aphelion and takes 12,050 years to complete. Mike Brown, who discovered the object in 2003, asserts that it cannot be part of the scattered disc or the Kuiper belt as its perihelion is too distant to have been affected by Neptune's migration. He and other astronomers consider it to be the first in an entirely new population, which also may include the object 2000 $CR_{105}$, which has a perihelion of 45 AU, an aphelion of 415 AU, and an orbital period of 3,420 years.[113] Brown terms this population the "inner Oort cloud", as it may have formed through a similar process, although it is far closer to the Sun.[114] Sedna is very likely a dwarf planet, though its shape has yet to be determined.

### Oort cloud

The Oort cloud is a hypothetical spherical cloud of up to a trillion icy objects that is believed to be the source for all long-period comets and to surround the Solar System at roughly 50,000 AU (around 1 light-year (ly)), and possibly to as far as 100,000 AU (1.87 ly). It is believed to be composed of comets that were ejected from the inner Solar System by gravitational interactions with the outer planets. Oort cloud objects move very slowly, and can be perturbed by infrequent events such as collisions, the gravitational effects of a passing star, or the galactic tide, the tidal force exerted by the Milky Way.[115] [116]

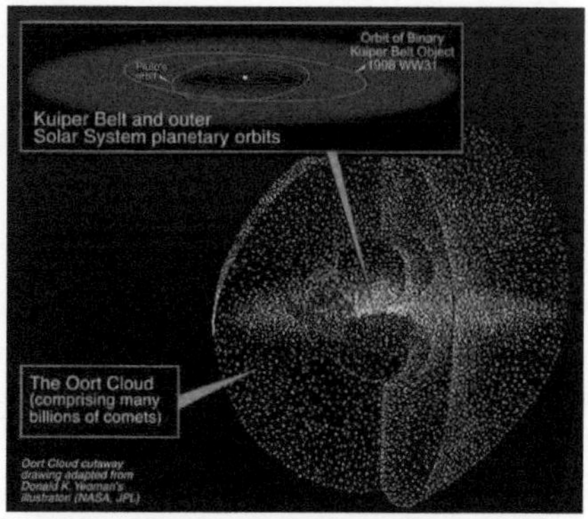

An artist's rendering of the Oort cloud, the Hills cloud, and the Kuiper belt (inset)

### Boundaries

Much of the Solar System is still unknown. The Sun's gravitational field is estimated to dominate the gravitational forces of surrounding stars out to about two light years (125,000 AU). Lower estimates for the radius of the Oort cloud, by contrast, do not place it farther than 50,000 AU.[117] Despite discoveries such as Sedna, the region between the Kuiper belt and the Oort cloud, an area tens of thousands of AU in radius, is still virtually unmapped. There are also ongoing studies of the region between Mercury and the Sun.[118] Objects may yet be discovered in the Solar System's uncharted regions.

In November 2012 NASA announced that as Voyager I approached the transition zone to the outer limit of the Solar System its instruments detected a sharp intensification of the magnetic field. No change in the direction of the magnetic field had occurred, which NASA scientists then interpreted to indicate that Voyager I had not yet left the Solar System.[119]

## Galactic context

The Solar System is located in the Milky Way galaxy, a barred spiral galaxy with a diameter of about 100,000 light-years containing about 200 billion stars.[120] The Sun resides in one of the Milky Way's outer spiral arms, known as the Orion–Cygnus Arm or Local Spur.[121] The Sun lies between 25,000 and 28,000 light years from the Galactic Centre,[122] and its speed within the galaxy is about 220 kilometres per second, so that it completes one revolution every 225–250 million years. This revolution is known as the Solar System's galactic year.[123] The solar apex, the direction of the Sun's path through interstellar space, is near the constellation of Hercules in the direction of the current location of the bright star Vega.[124] The plane of the ecliptic lies at an angle of about 60° to the galactic plane.[f]

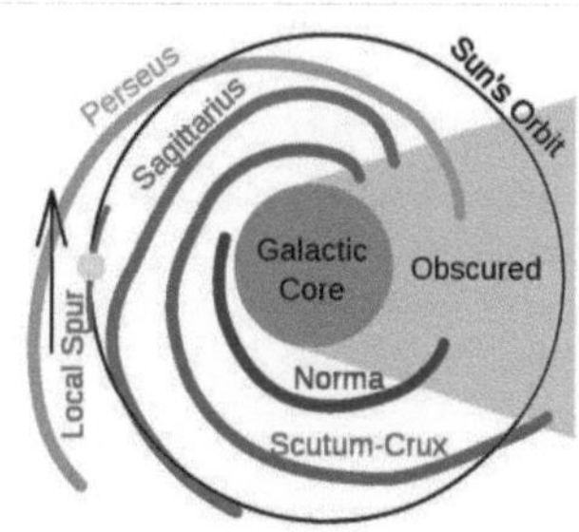

Location of the Solar System within our galaxy

The Solar System's location in the galaxy is a factor in the evolution of life on Earth. Its orbit is close to circular, and orbits near the Sun are at roughly the same speed as that of the spiral arms. Therefore, the Sun passes through arms only rarely. Since spiral arms are home to a far larger concentration of supernovae, gravitational instabilities, and radiation which could disrupt the Solar System, this has given Earth long periods of stability for life to evolve.[125]

The Solar System also lies well outside the star-crowded environs of the galactic centre. Near the centre, gravitational tugs from nearby stars could perturb bodies in the Oort Cloud and send many comets into the inner Solar System, producing collisions with potentially catastrophic implications for life on Earth. The intense radiation of the galactic centre could also interfere with the development of complex life.[125] Even at the Solar System's current location, some scientists have hypothesised that recent supernovae may have adversely affected life in the last 35,000 years by flinging pieces of expelled stellar core towards the Sun as radioactive dust grains and larger, comet-like bodies.[126]

### Neighbourhood

The immediate galactic neighbourhood of the Solar System is known as the Local Interstellar Cloud or Local Fluff, an area of denser cloud in an otherwise sparse region known as the Local Bubble, an hourglass-shaped cavity in the interstellar medium roughly 300 light years across. The bubble is suffused with high-temperature plasma that suggests it is the product of several recent supernovae.[127]

There are relatively few stars within ten light years (95 trillion km) of the Sun. The closest is the triple star system Alpha Centauri, which is about 4.4 light years away. Alpha Centauri A and B are a closely tied pair of Sun-like stars, while the small red dwarf Alpha Centauri C (also known as Proxima Centauri) orbits the pair at a distance of 0.2 light years. The stars next closest to the Sun are the red dwarfs Barnard's Star (at 5.9 light years), Wolf 359 (7.8 light years) and Lalande 21185 (8.3 light years). The largest star within ten light years is Sirius, a bright main-sequence star roughly twice the Sun's mass and orbited by a white dwarf called Sirius B. It lies 8.6 light years away. The remaining systems within ten light years are the binary red-dwarf system Luyten 726-8 (8.7 light years) and the solitary red dwarf Ross 154 (9.7 light years).[128] The Solar System's closest solitary sun-like star is Tau Ceti, which lies 11.9 light years away. It has roughly 80 percent the Sun's mass, but only 60 percent of its luminosity.[129] The closest known extrasolar planet to the Sun lies around Alpha Centauri B. Its one confirmed planet, Alpha Centauri Bb, is at least 1.1 times Earth's mass and orbits its star every 3.236 days.[130]

A diagram of our location in the observable Universe. (*Click here for an alternate image.*)

## Visual summary

A sampling of closely imaged Solar System bodies, selected for size and detail and sorted by volume. The Sun is approximately 10,000 times larger than, and 41 trillion times the volume of, the smallest object shown (Prometheus). Other lists include: List of Solar System objects by size, List of natural satellites, List of minor planets, and Lists of comets.

# Solar System

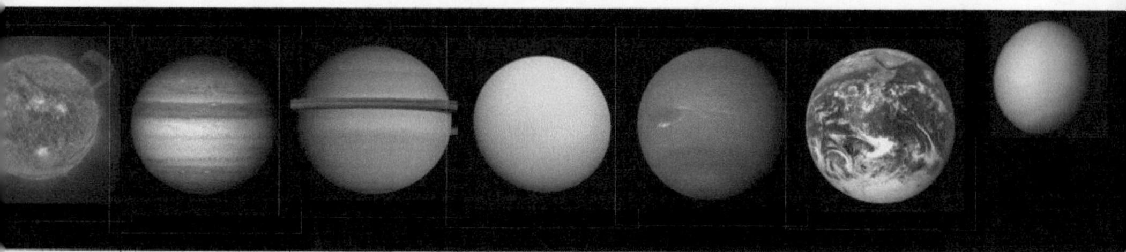

| Sun | Jupiter | Saturn | Uranus | Neptune | Earth | Venus |
|---|---|---|---|---|---|---|
| (star) | (planet) | (planet) | (planet) | (planet) | (planet) | (planet) |

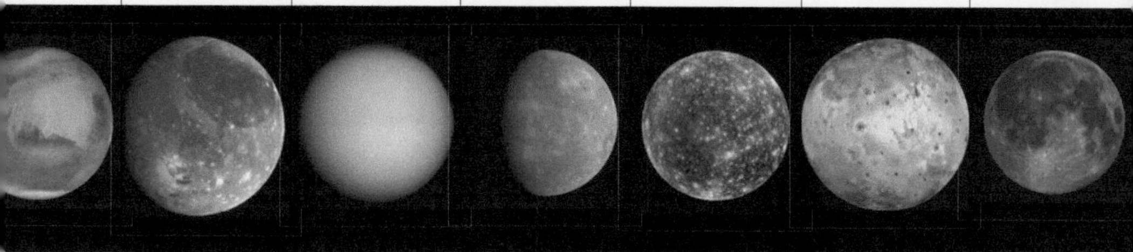

| Mars | Ganymede | Titan | Mercury | Callisto | Io | Moon |
|---|---|---|---|---|---|---|
| (planet) | (moon of Jupiter) | (moon of Saturn) | (planet) | (moon of Jupiter) | (moon of Jupiter) | (moon of Earth) |

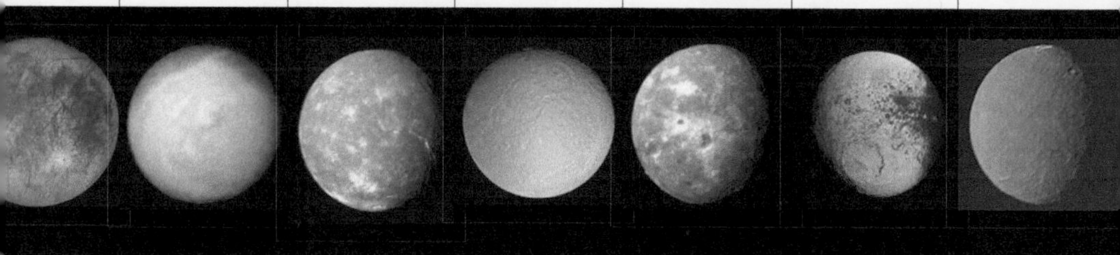

| Europa | Triton | Titania | Rhea | Oberon | Iapetus | Umbriel |
|---|---|---|---|---|---|---|
| on of Jupiter) | (moon of Neptune) | (moon of Uranus) | (moon of Saturn) | (moon of Uranus) | (moon of Saturn) | (moon of Uranus) |

| Ariel | Dione | Tethys | Vesta | Enceladus | Miranda | Proteus |
|---|---|---|---|---|---|---|
| on of Uranus) | (moon of Saturn) | (moon of Saturn) | (asteroid) | (moon of Saturn) | (moon of Uranus) | (moon of Neptune) |

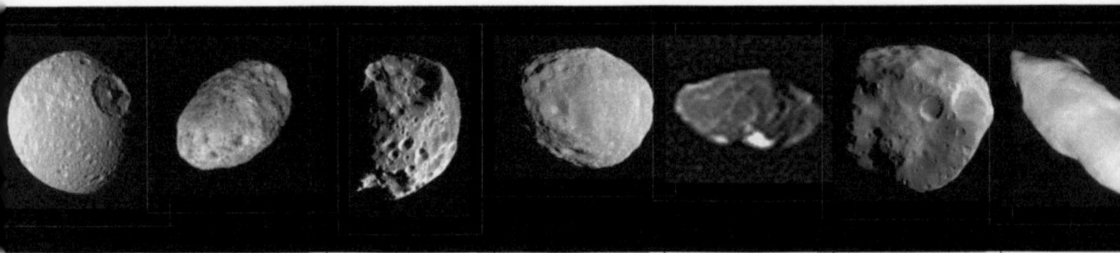

| Mimas (moon of Saturn) | Hyperion (moon of Saturn) | Phoebe (moon of Saturn) | Janus (moon of Saturn) | Amalthea (moon of Jupiter) | Epimetheus (moon of Saturn) | Prometheus (moon of Satu |
|---|---|---|---|---|---|---|

## See also

- Astronomical symbols
- List of gravitationally rounded objects of the Solar System
- List of geological features of the Solar System
- Planetary mnemonic
- Solar System in fiction

## Notes

a. Capitalization of the name varies. The IAU, the authoritative body regarding astronomical nomenclature, specifies capitalizing the names of all individual astronomical objects [131] (**Solar System**). However, the name is commonly rendered in lower case (**solar system**) – as, for example, in the *Oxford English Dictionary* and *Merriam-Webster's 11th Collegiate Dictionary* [132]

b. According to current definitions, objects in orbit around the Sun are classed dynamically and physically into three categories: *planets*, *dwarf planets*, and *small Solar System bodies*. A planet is any body in orbit around the Sun that has enough mass to form itself into a spherical shape and has cleared its immediate neighbourhood of all smaller objects. By this definition, the Solar System has eight known planets: Mercury, Venus, Earth, Mars, Jupiter, Saturn, Uranus, and Neptune. Pluto does not fit this definition, because it has not cleared its orbit of surrounding Kuiper belt objects.[133] A dwarf planet is a celestial body orbiting the Sun that is massive enough to be rounded by its own gravity but has not cleared its neighbouring region of planetesimals and is not a satellite.[133] The IAU has recognized five dwarf planets: Ceres, Pluto, Haumea, Makemake, and Eris.[97] However, only Ceres and Pluto have been observed in enough detail to prove that they fit the definition. Other objects may be accepted as dwarf planets in the future, such as Sedna, Orcus, and Quaoar.[134] Dwarf planets that orbit in the trans-Neptunian region are called "plutoids", though that term is not in widespread use.[135] The remainder of the objects in orbit around the Sun are small Solar System bodies.[133]

c. See List of natural satellites for the full list of natural satellites of the eight planets and first five dwarf planets.

d. The mass of the Solar System excluding the Sun, Jupiter and Saturn can be determined by adding together all the calculated masses for its largest objects and using rough calculations for the masses of the Oort cloud (estimated at roughly 3 Earth masses),[136] the Kuiper belt (estimated at roughly 0.1 Earth mass)[91] and the asteroid belt (estimated to be 0.0005 Earth mass)[73] for a total, rounded upwards, of ~37 Earth masses, or 8.1 percent the mass in orbit around the Sun. With the combined masses of Uranus and Neptune (~31 Earth masses) subtracted, the remaining ~6 Earth masses of material comprise 1.3 percent of the total.

e. Astronomers measure distances within the Solar System in astronomical units (AU). One AU equals the average distance between the centers of Earth and the Sun, or 149,598,000 km. Pluto is about 38 AU from the Sun and Jupiter is about 5.2 AU from the Sun. One light-year is 63,240 AU.

f. If $\psi$ is the angle between the north pole of the ecliptic and the north galactic pole then:

,

where 27° 07′ 42.01″ and 12h 51m 26.282 are the declination and right ascension of the north galactic pole,[137] while 66° 33′ 38.6″ and 18h 0m 00 are those for the north pole of the ecliptic. (Both pairs of coordinates are for J2000 epoch.) The result of the calculation is 60.19°.

## References


[1] Mike Brown (August 23, 2011). ""Free the dwarf planets!"" (http://www.mikebrownsplanets.com/2011/08/free-dwarf-planets.html). *"Mike Brown's Planets (self-published)".* .

[2] "How Many Solar System Bodies" (http://ssd.jpl.nasa.gov/?body_count). NASA/JPL Solar System Dynamics. . Retrieved 2012-12-13.

[3] Wm. Robert Johnston (2012-10-28). "Asteroids with Satellites" (http://www.johnstonsarchive.net/astro/asteroidmoons.html). Johnston's Archive. . Retrieved 2012-12-13.

[4] Error: Bad DOI specified!

[5] "Today we know of more than a dozen dwarf planets in the solar system". The PI's Perspective (http://pluto.jhuapl.edu/overview/piPerspective.php?page=piPerspective_08_24_2012)

[6] WC Rufus (1923). "The astronomical system of Copernicus". *Popular Astronomy* **31**: 510. Bibcode 1923PA.....31..510R.

[7] Weinert, Friedel (2009). *Copernicus, Darwin, & Freud: revolutions in the history and philosophy of science*. Wiley-Blackwell. p. 21. ISBN 978-1-4051-8183-9.

[8] M Woolfson (2000). "The origin and evolution of the solar system". *Astronomy & Geophysics* **41** (1): 1.12. doi:10.1046/j.1468-4004.2000.00012.x.

[9] Levison, H. F.; Morbidelli, A. (2003-11-27). "The formation of the Kuiper belt by the outward transport of bodies during Neptune's migration" (http://www.nature.com/nature/journal/v426/n6965/abs/nature02120.html). *Nature* **426**: 419–421. doi:10.1038/nature02120. PMID 14647375. . Retrieved 2012-05-26.

[10] Harold F. Levison, Martin J Duncan (1997). "From the Kuiper Belt to Jupiter-Family Comets: The Spatial Distribution of Ecliptic Comets". *Icarus* **127** (1): 13–32. Bibcode 1997Icar..127...13L. doi:10.1006/icar.1996.5637.

[11] Grossman, Lisa (13 August 2009). "Planet found orbiting its star backwards for first time" (http://www.newscientist.com/article/dn17603-planet-found-orbiting-its-star-backwards-for-first-time.html). NewScientist. . Retrieved 10 October 2009.

[12] nineplanets.org. "An Overview of the Solar System" (http://www.nineplanets.org/overview.html). . Retrieved 2007-02-15.

[13] Amir Alexander (2006). "New Horizons Set to Launch on 9-Year Voyage to Pluto and the Kuiper Belt" (http://www.planetary.org/news/2006/0116_New_Horizons_Set_to_Launch_on_9_Year.html). *The Planetary Society*. . Retrieved 2006-11-08.

[14] Marochnik, L. and Mukhin, L. (1995). ""Is Solar System Evolution Cometary Dominated?" (http://adsabs.harvard.edu/abs/1995ASPC...74...83M). In Shostak, G. S.. *Progress in the Search for Extraterrestrial Life*. Astronomical Society of the Pacific Conference Series. **74**. pp. 83. .

[15] Error: Bad DOI specified!

[16] "Dawn: A Journey to the Beginning of the Solar System" (http://www-ssc.igpp.ucla.edu/dawn/background.html). *Space Physics Center: UCLA*. 2005. . Retrieved 2007-11-03.

[17] "The Sun's Vital Statistics" (http://solar-center.stanford.edu/vitalstats.html). Stanford Solar Center. . Retrieved 2008-07-29., citing Eddy, J. (1979). *A New Sun: The Solar Results From Skylab* (http://history.nasa.gov/SP-402/contents.htm). NASA. p. 37. NASA SP-402. .

[18] Williams, Dr. David R. (September 7, 2006). "Saturn Fact Sheet" (http://nssdc.gsfc.nasa.gov/planetary/factsheet/saturnfact.html). NASA. . Retrieved 2007-07-31.

[19] Williams, Dr. David R. (November 16, 2004). "Jupiter Fact Sheet" (http://nssdc.gsfc.nasa.gov/planetary/factsheet/jupiterfact.html). NASA. . Retrieved 2007-08-08.

[20] Paul Robert Weissman, Torrence V. Johnson (2007). *Encyclopedia of the solar system*. Academic Press. p. 615. ISBN 0-12-088589-1.

[21] Error: Bad DOI specified!

[22] Error: Bad DOI specified!

[23] Michael Zellik (2002). *Astronomy: The Evolving Universe* (9th ed.). Cambridge University Press. p. 240. ISBN 0-521-80090-0. OCLC 223304585 46685453.

[24] Placxo, Kevin W.; Gross, Michael (2006). *Astrobiology: a brief introduction* (http://books.google.com/?id=2JuGDL144BEC&pg=PA66&dq=inventory+volatiles+hydrogen&q=inventory volatiles hydrogen). JHU Press. p. 66. ISBN 978-0-8018-8367-5. .

[25] Guy Ottewell (1989). "The Thousand-Yard Model" (http://www.noao.edu/education/peppercorn/pcmain.html). *NOAO Educational Outreach Office*. . Retrieved 2012-05-10.

[26] "Tours of Model Solar Systems" (http://internal.psychology.illinois.edu/~wbrewer/solarmodel.html). University of Illinois. . Retrieved 2012-05-10.

[27] "Luleå är Sedna. I alla fall om vår sol motsvaras av Globen i Stockholm." (http://www.kuriren.nu/arkiv/2005/11/17/Lokalt/1510647/LuleÃ¥-Ã¤r-Sedna.aspx). Norrbotten Kuriren (in swedish). . Retrieved 2010-05-10.

[28] The date is based on the oldest inclusions found to date in meteorites, and is thought to be the date of the formation of the first solid material in the collapsing nebula.

A. Bouvier and M. Wadhwa. "The age of the solar system redefined by the oldest Pb-Pb age of a meteoritic inclusion." *Nature Geoscience*, in press, 2010. doi:10.1038/NGEO941

[29] "Lecture 13: The Nebular Theory of the origin of the Solar System" (http://atropos.as.arizona.edu/aiz/teaching/nats102/mario/solar_system.html). *University of Arizona*. . Retrieved 2006-12-27.

[30] Irvine, W. M. (1983). "The chemical composition of the pre-solar nebula". *Cometary exploration; Proceedings of the International Conference*. **1**. pp. 3. Bibcode 1983coex....1....3I.

[31] Greaves, Jane S. (2005-01-07). "Disks Around Stars and the Growth of Planetary Systems". *Science* **307** (5706): 68–71. Bibcode 2005Sci...307...68G. doi:10.1126/science.1101979. PMID 15637266.

[32] "Present Understanding of the Origin of Planetary Systems" (http://www.nap.edu/openbook.php?record_id=1732&page=21). National Academy of Sciences. 2000-04-05. . Retrieved 2007-01-19.

[33] Boss, A. P.; Durisen, R. H. (2005). "Chondrule-forming Shock Fronts in the Solar Nebula: A Possible Unified Scenario for Planet and Chondrite Formation". *The Astrophysical Journal* **621** (2): L137. Bibcode 2005ApJ...621L.137B. doi:10.1086/429160.

[34] Sukyoung Yi; Pierre Demarque; Yong-Cheol Kim; Young-Wook Lee; Chang H. Ree; Thibault Lejeune; Sydney Barnes (2001). "Toward Better Age Estimates for Stellar Populations: The Isochrones for Solar Mixture". *Astrophysical Journal Supplement* **136**: 417. arXiv:astro-ph/0104292. Bibcode 2001ApJS..136..417Y. doi:10.1086/321795.

[35] A. Chrysostomou, P. W. Lucas (2005). "The Formation of Stars". *Contemporary Physics* **46** (1): 29. Bibcode 2005ConPh..46...29C. doi:10.1080/0010751042000275277.

[36] K. P. Schroder, Robert Cannon Smith (2008). "Distant future of the Sun and Earth revisited". *Monthly Notices of the Royal Astronomical Society* **386** (1): 155–163. Bibcode 2008MNRAS.386..155S. doi:10.1111/j.1365-2966.2008.13022.x.

[37] Pogge, Richard W. (1997). "The Once & Future Sun" (http://web.archive.org/web/20050527094435/http://www-astronomy.mps.ohio-state.edu/Vistas/) (lecture notes). *New Vistas in Astronomy* (http://www-astronomy.mps.ohio-state.edu/Vistas/). Archived from the original (http://www.astronomy.ohio-state.edu/~pogge/Lectures/vistas97.html) on May 27, 2005. . Retrieved 2005-12-07.

[38] "Sun: Facts & Figures" (http://web.archive.org/web/20080102034758/http://solarsystem.nasa.gov/planets/profile.cfm?Object=Sun&Display=Facts&System=Metric). NASA. Archived from the original (http://solarsystem.nasa.gov/planets/profile.cfm?Object=Sun&Display=Facts&System=Metric) on 2008-01-02. . Retrieved 2009-05-14.

[39] Zirker, Jack B. (2002). *Journey from the Center of the Sun*. Princeton University Press. pp. 120–127. ISBN 978-0-691-05781-1.

[40] "Why is visible light visible, but not other parts of the spectrum?" (http://www.straightdope.com/columns/read/2085/why-is-visible-light-visible-but-not-other-parts-of-the-spectrum). The Straight Dome. 2003. . Retrieved 2009-05-14.

[41] Than, Ker (January 30, 2006). "Astronomers Had it Wrong: Most Stars are Single" (http://www.space.com/scienceastronomy/060130_mm_single_stars.html). SPACE.com. . Retrieved 2007-08-01.

[42] Smart, R. L.; Carollo, D.; Lattanzi, M. G.; McLean, B.; Spagna, A. (2001). "The Second Guide Star Catalogue and Cool Stars". In Hugh R. A. Jones and Iain A. Steele. *Ultracool Dwarfs: New Spectral Types L and T*. Springer. pp. 119. Bibcode 2001udns.conf..119S.

[43] Nir J. Shaviv (2003). "Towards a Solution to the Early Faint Sun Paradox: A Lower Cosmic Ray Flux from a Stronger Solar Wind". *Journal of Geophysical Research* **108** (A12): 1437. arXiv:astroph/0306477. Bibcode 2003JGRA..108.1437S. doi:10.1029/2003JA009997.

[44] T. S. van Albada, Norman Baker (1973). "On the Two Oosterhoff Groups of Globular Clusters". *Astrophysical Journal* **185**: 477–498. Bibcode 1973ApJ...185..477V. doi:10.1086/152434.

[45] Charles H. Lineweaver (2001-03-09). "An Estimate of the Age Distribution of Terrestrial Planets in the Universe: Quantifying Metallicity as a Selection Effect". *Icarus* **151** (2): 307–313. Bibcode 2001Icar..151..307L. doi:10.1006/icar.2001.6607.

[46] "Solar Physics: The Solar Wind" (http://solarscience.msfc.nasa.gov/SolarWind.shtml). *Marshall Space Flight Center*. 2006-07-16. . Retrieved 2006-10-03.

[47] "Voyager Enters Solar System's Final Frontier" (http://www.nasa.gov/vision/universe/solarsystem/voyager_agu.html). *NASA*. . Retrieved 2007-04-02.

[48] Phillips, Tony (2001-02-15). "The Sun Does a Flip" (http://science.nasa.gov/headlines/y2001/ast15feb_1.htm). *Science@NASA*. . Retrieved 2007-02-04.

[49] A Star with two North Poles (http://science.nasa.gov/headlines/y2003/22apr_currentsheet.htm), April 22, 2003, Science @ NASA

[50] Riley, Pete (2002). "Modeling the heliospheric current sheet: Solar cycle variations" (http://ulysses.jpl.nasa.gov/science/monthly_highlights/2002-July-2001JA000299.pdf). *Journal of Geophysical Research* **107**. Bibcode 2002JGRA.107g.SSH8R. doi:10.1029/2001JA000299. .

[51] Lundin, Richard (2001-03-09). "Erosion by the Solar Wind". *Science* **291** (5510): 1909. doi:10.1126/science.1059763. PMID 11245195.

[52] Langner, U. W.; M. S. Potgieter (2005). "Effects of the position of the solar wind termination shock and the heliopause on the heliospheric modulation of cosmic rays". *Advances in Space Research* **35** (12): 2084–2090. Bibcode 2005AdSpR..35.2084L. doi:10.1016/j.asr.2004.12.005.

[53] "Long-term Evolution of the Zodiacal Cloud" (http://astrobiology.arc.nasa.gov/workshops/1997/zodiac/backman/IIIc.html). 1998. . Retrieved 2007-02-03.

[54] "ESA scientist discovers a way to shortlist stars that might have planets" (http://sci.esa.int/science-e/www/object/index.cfm?fobjectid=29471). *ESA Science and Technology*. 2003. . Retrieved 2007-02-03.

[55] Landgraf, M.; Liou, J.-C.; Zook, H. A.; Grün, E. (May 2002). "Origins of Solar System Dust beyond Jupiter" (http://astron.berkeley.edu/~kalas/disksite/library/ladgraf02.pdf). *The Astronomical Journal* **123** (5): 2857–2861. Bibcode 2002AJ....123.2857L. doi:10.1086/339704. . Retrieved 2007-02-09.

[56] "Inner Solar System" (http://nasascience.nasa.gov/planetary-science/exploring-the-inner-solar-system). NASA Science (Planets). . Retrieved 2009-05-09.

[57] Schenk P., Melosh H. J. (1994), *Lobate Thrust Scarps and the Thickness of Mercury's Lithosphere*, Abstracts of the 25th Lunar and Planetary Science Conference, 1994LPI....25.1203S

[58] Bill Arnett (2006). "Mercury" (http://www.nineplanets.org/mercury.html). *The Nine Planets*. . Retrieved 2006-09-14.

[59] Benz, W., Slattery, W. L., Cameron, A. G. W. (1988), *Collisional stripping of Mercury's mantle*, Icarus, v. 74, p. 516–528.

[60] Cameron, A. G. W. (1985), *The partial volatilization of Mercury*, Icarus, v. 64, p. 285–294.

[61] Mark Alan Bullock (1997) (PDF). *The Stability of Climate on Venus* (http://www.boulder.swri.edu/~bullock/Homedocs/PhDThesis.pdf). Southwest Research Institute. . Retrieved 2006-12-26.

[62] Paul Rincon (1999). "Climate Change as a Regulator of Tectonics on Venus" (http://www.boulder.swri.edu/~bullock/Homedocs/Science2_1999.pdf) (PDF). *Johnson Space Center Houston, TX, Institute of Meteoritics, University of New Mexico, Albuquerque, NM*. . Retrieved 2006-11-19.

[63] "What are the characteristics of the Solar System that lead to the origins of life?" (http://science.nasa.gov/planetary-science/big-questions/what-are-the-characteristics-of-the-solar-system-that-lead-to-the-origins-of-life-1/). NASA Science (Big Questions). . Retrieved 2011-08-30.

[64] Anne E. Egger, M.A./M.S.. "Earth's Atmosphere: Composition and Structure" (http://www.visionlearning.com/library/module_viewer.php?c3=&mid=107&l=). *VisionLearning.com*. . Retrieved 2006-12-26.

[65] David C. Gatling, Conway Leovy (2007). "Mars Atmosphere: History and Surface Interactions". In Lucy-Ann McFadden et. al.. *Encyclopaedia of the Solar System*. pp. 301–314.

[66] David Noever (2004). "Modern Martian Marvels: Volcanoes?" (http://www.astrobio.net/news/modules.php?op=modload&name=News&file=article&sid=1360&mode=thread&order=0&thold=0). *NASA Astrobiology Magazine*. . Retrieved 2006-07-23.

[67] "Mars: A Kid's Eye View" (http://solarsystem.nasa.gov/planets/profile.cfm?Object=Mars&Display=Kids). NASA. . Retrieved 2009-05-14.

[68] Scott S. Sheppard, David Jewitt, and Jan Kleyna (2004). "A Survey for Outer Satellites of Mars: Limits to Completeness" (http://www2.ess.ucla.edu/~jewitt/papers/2004/SJK2004.pdf). *Astronomical Journal*. . Retrieved 2006-12-26.

[69] "Are Kuiper Belt Objects asteroids? Are large Kuiper Belt Objects planets?" (http://curious.astro.cornell.edu/question.php?number=601). Cornell University. . Retrieved 2009-03-01.

[70] Petit, J.-M.; Morbidelli, A.; Chambers, J. (2001). "The Primordial Excitation and Clearing of the Asteroid Belt" (http://www.gps.caltech.edu/classes/ge133/reading/asteroids.pdf) (PDF). *Icarus* **153** (2): 338–347. Bibcode 2001Icar..153..338P. doi:10.1006/icar.2001.6702. . Retrieved 2007-03-22.

[71] "IAU Planet Definition Committee" (http://www.iau.org/public_press/news/release/iau0601/newspaper/). International Astronomical Union. 2006. . Retrieved 2009-03-01.

[72] "New study reveals twice as many asteroids as previously believed" (http://www.esa.int/esaCP/ESAASPF18ZC_index_0.html). *ESA*. 2002. . Retrieved 2006-06-23.

[73] Krasinsky, G. A.; Pitjeva, E. V.; Vasilyev, M. V.; Yagudina, E. I. (July 2002). "Hidden Mass in the Asteroid Belt". *Icarus* **158** (1): 98–105. Bibcode 2002Icar..158...98K. doi:10.1006/icar.2002.6837.

[74] Beech, M.; Steel; Duncan I. Steel (September 1995). "On the Definition of the Term Meteoroid". *Quarterly Journal of the Royal Astronomical Society* **36** (3): 281–284. Bibcode 1995QJRAS..36..281B.

[75] "History and Discovery of Asteroids" (http://dawn.jpl.nasa.gov/DawnClassrooms/1_hist_dawn/history_discovery/Development/a_modeling_scale.doc) (DOC). *NASA*. . Retrieved 2006-08-29.

[76] Phil Berardelli (2006). "Main-Belt Comets May Have Been Source Of Earths Water" (http://www.spacedaily.com/reports/Main_Belt_Comets_May_Have_Been_Source_Of_Earths_Water.html). *SpaceDaily*. . Retrieved 2006-06-23.

[77] Barucci, M. A.; Kruikshank, D.P.; Mottola S.; Lazzarin M. (2002). "Physical Properties of Trojan and Centaur Asteroids". *Asteroids III*. Tucson, Arizona: University of Arizona Press. pp. 273–87.

[78] A. Morbidelli, W. F. Bottke Jr., Ch. Froeschlé, P. Michel; Bottke; Froeschlé; Michel (January 2002). W. F. Bottke Jr., A. Cellino, P. Paolicchi, and R. P. Binzel. ed. "Origin and Evolution of Near-Earth Objects" (http://www.boulder.swri.edu/~bottke/Reprints/Morbidelli-etal_2002_AstIII_NEOs.pdf) (PDF). *Asteroids III* (University of Arizona Press): 409–422. Bibcode 2002aste.conf..409M. .

[79] Jack J. Lissauer, David J. Stevenson (2006). "Formation of Giant Planets" (http://www.gps.caltech.edu/uploads/File/People/djs/lissauer&stevenson(PPV).pdf) (PDF). *NASA Ames Research Center; California Institute of Technology*. . Retrieved 2006-01-16.

[80] Pappalardo, R T (1999). "Geology of the Icy Galilean Satellites: A Framework for Compositional Studies" (http://www.agu.org/cgi-bin/SFgate/SFgate?&listenv=table&multiple=1&range=1&directget=1&application=fm99&database=/data/epubs/wais/indexes/fm99/fm99&maxhits=200&="P11C-10"). *Brown University*. . Retrieved 2006-01-16.

[81] Kargel, J. S. (1994). "Cryovolcanism on the icy satellites". *Earth, Moon, and Planets* **67**: 101–113. Bibcode 1995EM&P...67..101K. doi:10.1007/BF00613296.

[82] Hawksett, David; Longstaff, Alan; Cooper, Keith; Clark, Stuart; Longstaff; Cooper; Clark (2005). "10 Mysteries of the Solar System". *Astronomy Now* **19**: 65. Bibcode 2005AsNow..19h..65H.

[83] Podolak, M.; Reynolds, R. T.; Young, R. (1990). "Post Voyager comparisons of the interiors of Uranus and Neptune". *NASA, Ames Research Center* **17** (10): 1737. Bibcode 1990GeoRL..17.1737P. doi:10.1029/GL017i010p01737.

[84] Duxbury, N. S., Brown, R. H. (1995). "The Plausibility of Boiling Geysers on Triton" (http://trs-new.jpl.nasa.gov/dspace/handle/2014/28034?mode=full). *Beacon eSpace*. . Retrieved 2006-01-16.

[85] John Stansberry, Will Grundy, Mike Brown, Dale Cruikshank, John Spencer, David Trilling, Jean-Luc Margot (2007). "Physical Properties of Kuiper Belt and Centaur Objects: Constraints from Spitzer Space Telescope". *The Solar System Beyond Neptune*. pp. 161. arXiv:astro-ph/0702538. Bibcode 2008ssbn.book..161S.

[86] Patrick Vanouplines (1995). "Chiron biography" (http://www.vub.ac.be/STER/www.astro/chibio.htm). *Vrije Universitiet Brussel*. . Retrieved 2006-06-23.

[87] Sekanina, Zdeněk (2001). "Kreutz sungrazers: the ultimate case of cometary fragmentation and disintegration?". *Publications of the Astronomical Institute of the Academy of Sciences of the Czech Republic* **89**: 78–93. Bibcode 2001PAICz..89...78S.

[88] Królikowska, M. (2001). "A study of the original orbits of *hyperbolic* comets". *Astronomy & Astrophysics* **376** (1): 316–324. Bibcode 2001A&A...376..316K. doi:10.1051/0004-6361:20010945.

[89] Whipple, Fred L. (1992). "The activities of comets related to their aging and origin". *Celestial Mechanics and Dynamical Astronomy* **54**: 1–11. Bibcode 1992CeMDA..54....1W. doi:10.1007/BF00049540.

[90] Stephen C. Tegler (2007). "Kuiper Belt Objects: Physical Studies". In Lucy-Ann McFadden et. al.. *Encyclopedia of the Solar System*. pp. 605–620.

[91] Audrey Delsanti and David Jewitt (2006). "The Solar System Beyond The Planets" (http://web.archive.org/web/20070129151907/http://www.ifa.hawaii.edu/faculty/jewitt/papers/2006/DJ06.pdf) (PDF). *Institute for Astronomy, University of Hawaii*. Archived from the original (http://www.ifa.hawaii.edu/faculty/jewitt/papers/2006/DJ06.pdf) on January 29, 2007. . Retrieved 2007-01-03.

[92] Error: Bad DOI specified!

[93] Chiang *et al.*; Jordan, A. B.; Millis, R. L.; Buie, M. W.; Wasserman, L. H.; Elliot, J. L.; Kern, S. D.; Trilling, D. E. et al. (2003). "Resonance Occupation in the Kuiper Belt: Case Examples of the 5:2 and Trojan Resonances" (http://www.boulder.swri.edu/~buie/biblio/pub047.pdf). *The Astronomical Journal* **126** (1): 430–443. Bibcode 2003AJ....126..430C. doi:10.1086/375207. . Retrieved 2009-08-15.

[94] M. W. Buie, R. L. Millis, L. H. Wasserman, J. L. Elliot, S. D. Kern, K. B. Clancy, E. I. Chiang, A. B. Jordan, K. J. Meech, R. M. Wagner, D. E. Trilling (2005). "Procedures, Resources and Selected Results of the Deep Ecliptic Survey". *Earth, Moon, and Planets* **92** (1): 113. arXiv:astro-ph/0309251. Bibcode 2003EM&P...92..113B. doi:10.1023/B:MOON.0000031930.13823.be.

[95] E. Dotto1, M. A. Barucci2, and M. Fulchignoni (2006-08-24). "Beyond Neptune, the new frontier of the Solar System" (http://sait.oat.ts.astro.it/MSAIS/3/PDF/20.pdf) (PDF). . Retrieved 2006-12-26.

[96] Fajans, J.; L. Frièdland (October 2001). "Autoresonant (nonstationary) excitation of pendulums, Plutinos, plasmas, and other nonlinear oscillators" (http://ist-socrates.berkeley.edu/~fajans/pub/pdffiles/AutoPendAJP.pdf). *American Journal of Physics* **69** (10): 1096–1102. doi:10.1119/1.1389278. . Retrieved 2006-12-26.

[97] "Dwarf Planets and their Systems" (http://planetarynames.wr.usgs.gov/append7.html#DwarfPlanets). *Working Group for Planetary System Nomenclature (WGPSN)*. U.S. Geological Survey. 2008-11-07. . Retrieved 2008-07-13.

[98] Marc W. Buie (2008-04-05). "Orbit Fit and Astrometric record for 136472" (http://www.boulder.swri.edu/~buie/kbo/astrom/136472.html). SwRI (Space Science Department). . Retrieved 2012-07-15.

[99] Michael E. Brown. "The largest Kuiper belt objects" (http://www.gps.caltech.edu/~mbrown/papers/ps/kbochap.pdf) (PDF). *CalTech*. . Retrieved 2012-07-15.

[100] "News Release – IAU0807: IAU names fifth dwarf planet Haumea" (http://www.iau.org/public_press/news/release/iau0807/). *International Astronomical Union*. 2008-09-17. . Retrieved 2012-07-15.

[101] David Jewitt (2005). "The 1000 km Scale KBOs" (http://www2.ess.ucla.edu/~jewitt/kb/big_kbo.html). *University of Hawaii*. . Retrieved 2006-07-16.

[102] "List Of Centaurs and Scattered-Disk Objects" (http://www.minorplanetcenter.org/iau/lists/Centaurs.html). *IAU: Minor Planet Center*. . Retrieved 2007-04-02.

[103] Error: Bad DOI specified!

[104] Littmann, Mark (2004). *Planets Beyond: Discovering the Outer Solar System*. Courier Dover Publications. pp. 162–163. ISBN 978-0-486-43602-9.

[105] Fahr, H. J.; Kausch, T.; Scherer, H.; Kausch; Scherer (2000). "A 5-fluid hydrodynamic approach to model the Solar System-interstellar medium interaction" (http://aa.springer.de/papers/0357001/2300268.pdf) (PDF). *Astronomy & Astrophysics* **357**: 268. Bibcode 2000A&A...357..268F. . See Figures 1 and 2.

[106] NASA/JPL (2009). "Cassini's Big Sky: The View from the Center of Our Solar System" (http://www.jpl.nasa.gov/news/features.cfm?feature=2370&msource=F20091119&tr=y&auid=5615216). . Retrieved 2009-12-20.

[107] Stone, E. C.; Cummings, A. C.; McDonald, F. B.; Heikkila, B. C.; Lal, N.; Webber, W. R. (September 2005). "Voyager 1 explores the termination shock region and the heliosheath beyond". *Science* **309** (5743): 2017–20. Bibcode 2005Sci...309.2017S. doi:10.1126/science.1117684. PMID 16179468.

[108] Stone, E. C.; Cummings, A. C.; McDonald, F. B.; Heikkila, B. C.; Lal, N.; Webber, W. R. (July 2008). "An asymmetric solar wind termination shock". *Nature* **454** (7200): 71–4. doi:10.1038/nature07022. PMID 18596802.

[109] P. C. Frisch (University of Chicago) (June 24, 2002). "The Sun's Heliosphere & Heliopause" (http://antwrp.gsfc.nasa.gov/apod/ap020624.html). *Astronomy Picture of the Day*. . Retrieved 2006-06-23.

[110] "Voyager: Interstellar Mission" (http://voyager.jpl.nasa.gov/mission/interstellar.html). *NASA Jet Propulsion Laboratory*. 2007. . Retrieved 2008-05-08.

[111] R. L. McNutt, Jr. et al. (2006). "Innovative Interstellar Explorer". *Physics of the Inner Heliosheath: Voyager Observations, Theory, and Future Prospects*. AIP Conference Proceedings. **858**. pp. 341–347. Bibcode 2006AIPC..858..341M. doi:10.1063/1.2359348.

[112] Anderson, Mark (2007-01-05). "Interstellar space, and step on it!" (http://space.newscientist.com/article/mg19325850.900-interstellar-space-and-step-on-it.html). *New Scientist*. . Retrieved 2007-02-05.

[113] David Jewitt (2004). "Sedna – 2003 $VB_{12}$" (http://www2.ess.ucla.edu/~jewitt/kb/sedna.html). *University of Hawaii*. . Retrieved 2006-06-23.

[114] Mike Brown. "Sedna" (http://www.gps.caltech.edu/~mbrown/sedna/). *CalTech*. . Retrieved 2007-05-02.

[115] Stern SA, Weissman PR. (2001). "Rapid collisional evolution of comets during the formation of the Oort cloud." (http://www.ncbi.nlm.nih.gov/entrez/query.fcgi?cmd=Retrieve&db=PubMed&list_uids=11214311&dopt=Citation). *Space Studies Department, Southwest Research Institute, Boulder, Colorado*. . Retrieved 2006-11-19.

[116] Bill Arnett (2006). "The Kuiper Belt and the Oort Cloud" (http://www.nineplanets.org/kboc.html). *nineplanets.org*. . Retrieved 2006-06-23.

[117] T. Encrenaz, JP. Bibring, M. Blanc, MA. Barucci, F. Roques, PH. Zarka (2004). *The Solar System: Third edition*. Springer. p. 1.

[118] Durda D. D.; Stern S. A.; Colwell W. B.; Parker J. W.; Levison H. F.; Hassler D. M. (2004). "A New Observational Search for Vulcanoids in SOHO/LASCO Coronagraph Images". *Icarus* **148**: 312–315. Bibcode 2000Icar..148..312D. doi:10.1006/icar.2000.6520.

[119] http://www.nasa.gov/mission_pages/voyager/voyager20121203.html

[120] English, J. (2000). "Exposing the Stuff Between the Stars" (http://www.ras.ucalgary.ca/CGPS/press/aas00/pr/pr_14012000/pr_14012000map1.html) (Press release). Hubble News Desk. . Retrieved 2007-05-10.

[121] R. Drimmel, D. N. Spergel (2001). "Three Dimensional Structure of the Milky Way Disk". *Astrophysical Journal* **556**: 181–202. arXiv:astro-ph/0101259. Bibcode 2001ApJ...556..181D. doi:10.1086/321556.

[122] Eisenhauer, F.; et al. (2003). "A Geometric Determination of the Distance to the Galactic Center". *Astrophysical Journal* **597** (2): L121–L124. Bibcode 2003ApJ...597L.121E. doi:10.1086/380188.

[123] Leong, Stacy (2002). "Period of the Sun's Orbit around the Galaxy (Cosmic Year)" (http://hypertextbook.com/facts/2002/StacyLeong.shtml). *The Physics Factbook*. . Retrieved 2007-04-02.

[124] C. Barbieri (2003). "Elementi di Astronomia e Astrofisica per il Corso di Ingegneria Aerospaziale V settimana" (http://dipastro.pd.astro.it/planets/barbieri/Lezioni-AstroAstrofIng04_05-Prima-Settimana.ppt). *IdealStars.com*. . Retrieved 2007-02-12.

[125] Leslie Mullen (2001). "Galactic Habitable Zones" (http://www.astrobio.net/news/modules.php?op=modload&name=News&file=article&sid=139). *Astrobiology Magazine*. . Retrieved 2006-06-23.

[126] "Supernova Explosion May Have Caused Mammoth Extinction" (http://www.physorg.com/news6734.html). *Physorg.com*. 2005. . Retrieved 2007-02-02.

[127] "Near-Earth Supernovas" (http://science.nasa.gov/headlines/y2003/06jan_bubble.htm). *NASA*. . Retrieved 2006-07-23.

[128] "Stars within 10 light years" (http://www.solstation.com/stars/s10ly.htm). *SolStation*. . Retrieved 2007-04-02.

[129] "Tau Ceti" (http://www.solstation.com/stars/tau-ceti.htm). *SolStation*. . Retrieved 2007-04-02.

[130] Error: Bad DOI specified!

[131] http://www.iau.org/public_press/themes/naming/

[132] http://www.m-w.com/dictionary/solar%20system

[133] "The Final IAU Resolution on the definition of "planet" ready for voting" (http://www.iau.org/iau0602.423.0.html). IAU. 2006-08-24. . Retrieved 2007-03-02.

[134] Ron Ekers. "IAU Planet Definition Committee" (http://www.iau.org/public_press/news/release/iau0601/newspaper/). International Astronomical Union. . Retrieved 2008-10-13.

[135] "Plutoid chosen as name for Solar System objects like Pluto" (http://www.iau.org/public_press/news/release/iau0804). International Astronomical Union. June 11, 2008, Paris. . Retrieved 2008-06-11.

[136] Alessandro Morbidelli (2005). "Origin and dynamical evolution of comets and their reservoirs". arXiv:astro-ph/0512256.

[137] Reid, M.J.; Brunthaler, A. (2004 2004). "The Proper Motion of Sagittarius A*". *The Astrophysical Journal* **616** (2): 883. Bibcode 2004ApJ...616..872R. doi:10.1086/424960.

# External links

- Solar System Profile (http://solarsystem.nasa.gov/planets/profile.cfm?Object=SolarSys&Display=Overview) by NASA's Solar System Exploration (http://solarsystem.nasa.gov/index.cfm)
- NASA's Solar System Simulator (http://space.jpl.nasa.gov)
- NASA/JPL Solar System main page (http://www.jpl.nasa.gov/solar_system)

# Kuiper_belt

The **Kuiper belt** /ˈkaɪpər/ (rhymes with "viper"), sometimes called the **Edgeworth–Kuiper belt**, is a region of the Solar System beyond the planets, extending from the orbit of Neptune (at 30 AU) to approximately 50 AU from the Sun.[1] It is similar to the asteroid belt, but it is far larger—20 times as wide and 20 to 200 times as massive.[2] [3] Like the asteroid belt, it consists mainly of small bodies, or remnants from the Solar System's formation. While most asteroids are composed primarily of rock and metal, most Kuiper belt objects are composed largely of frozen volatiles (termed "ices"), such as methane, ammonia and water. The classical belt is home to at least three dwarf planets: Pluto, Haumea, and Makemake. Some of the Solar System's moons, such as Neptune's Triton and Saturn's Phoebe, are also believed to have originated in the region.[4] [5]

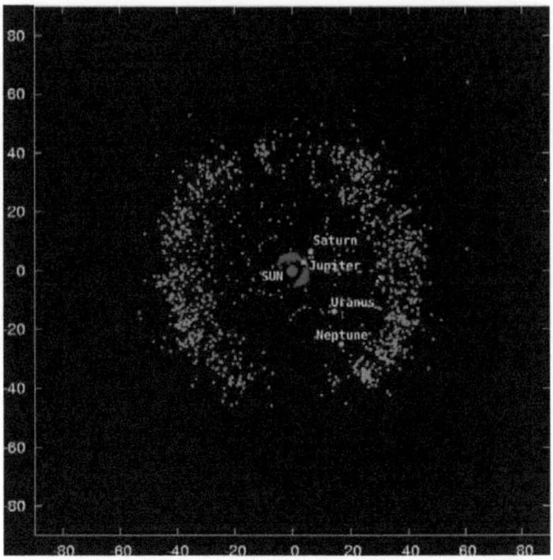

Known objects in the Kuiper belt, derived from data from the Minor Planet Center. Objects in the main belt are colored green, while scattered objects are colored orange. The four outer planets are blue. Neptune's few known trojans are yellow, while Jupiter's are pink. The scattered objects between Jupiter's orbit and the Kuiper belt are known as centaurs. The scale is in astronomical units. The pronounced gap at the bottom is due to difficulties in detection against the background of the plane of the Milky Way.

Since the belt was discovered in 1992,[6] the number of known Kuiper belt objects (KBOs) has increased to over a thousand, and more than 100,000 KBOs over 100 km (62 mi) in diameter are believed to exist.[7] The Kuiper belt was initially thought to be the main repository for periodic comets, those with orbits lasting less than 200 years. However, studies since the mid-1990s have shown that the classical belt is dynamically stable, and that comets' true place of origin is the scattered disc, a dynamically active zone created by the outward motion of Neptune 4.5 billion years ago;[8] scattered disc objects such as Eris have extremely eccentric orbits that take them as far as 100 AU from the Sun.[nb 1]

The Kuiper belt should not be confused with the hypothesized Oort cloud, which is a thousand times more distant. The objects within the Kuiper belt, together with the members of the scattered disc and any potential Hills cloud or Oort cloud objects, are collectively referred to as trans-Neptunian objects (TNOs).[9]

Pluto is the largest known member of the Kuiper belt, and the second largest known TNO, after the scattered-disc object Eris.[nb 1] Originally considered a planet, Pluto's status as part of the Kuiper belt caused it to be reclassified as a "dwarf planet" in 2006. It is compositionally similar to many other objects of the Kuiper belt, and its orbital period is characteristic of a class of KBOs known as "plutinos" which share the same 2:3 resonance with Neptune.

## History

Since the discovery of Pluto, many have speculated that it might not be alone. The region now called the Kuiper belt had been hypothesized in various forms for decades. It was only in 1992 that the first direct evidence for its existence was found. The number and variety of prior speculations on the nature of the Kuiper belt have led to continued uncertainty as to who deserves credit for first proposing it.

### Hypotheses

The first astronomer to suggest the existence of a trans-Neptunian population was Frederick C. Leonard. In 1930, soon after Pluto's discovery by Clyde Tombaugh, Leonard pondered whether it was "not likely that in Pluto there has come to light the *first* of a *series* of ultra-Neptunian bodies, the remaining members of which still await discovery but which are destined eventually to be detected".[10]

In 1943, in the *Journal of the British Astronomical Association*, Kenneth Edgeworth hypothesized that, in the region beyond Neptune, the material within the primordial solar nebula was too widely spaced to condense into planets, and so rather condensed into a myriad of smaller bodies. From this he concluded that "the outer region of the solar system, beyond the orbits of the planets, is occupied by a very large number of comparatively small bodies"[11] and that, from time to time, one of their number "wanders from its own sphere and appears as an occasional visitor to the inner solar system",[12] becoming a comet.

Astronomer Gerard Kuiper, after whom the Kuiper belt is named

In 1951, in an article for the journal *Astrophysics*, Gerard Kuiper speculated on a similar disc having formed early in the Solar System's evolution; however, he did not believe that such a belt still existed today. Kuiper was operating on the assumption common in his time, that Pluto was the size of the Earth, and had therefore scattered these bodies out toward the Oort cloud or out of the Solar System. Were Kuiper's hypothesis correct, there would not be a Kuiper belt today.[13]

The hypothesis took many other forms in the following decades. In 1962, physicist Al G.W. Cameron postulated the existence of "a tremendous mass of small material on the outskirts of the solar system",[14] while in 1964, Fred Whipple, who popularised the famous "dirty snowball" hypothesis for cometary structure, thought that a "comet belt" might be massive enough to cause the purported discrepancies in the orbit of Uranus that had sparked the search for Planet X, or at the very least, to affect the orbits of known comets.[15] Observation, however, ruled out this hypothesis.[14]

In 1977, Charles Kowal discovered 2060 Chiron, an icy planetoid with an orbit between Saturn and Uranus. He used a blink comparator; the same device that had allowed Clyde Tombaugh to discover Pluto nearly 50 years before.[16] In 1992, another object, 5145 Pholus, was discovered in a similar orbit.[17] Today, an entire population of comet-like bodies, the centaurs, is known to exist in the region between Jupiter and Neptune. The centaurs' orbits are unstable and have dynamical lifetimes of a few million years.[18] From the time of Chiron's discovery, astronomers speculated that they therefore must be frequently replenished by some outer reservoir.[19]

Further evidence for the belt's existence later emerged from the study of comets. That comets have finite lifespans has been known for some time. As they approach the Sun, its heat causes their volatile surfaces to sublimate into space, eating them gradually away. In order to still be visible over the age of the Solar System, they must be frequently replenished.[20] One such area of replenishment is the Oort cloud, the spherical swarm of comets extending beyond 50 000 AU from the Sun first hypothesised by astronomer Jan Oort in 1950.[21] It is believed to be the point of origin for long-period comets, those, like Hale–Bopp, with orbits lasting thousands of years.

There is however another comet population, known as short-period or periodic comets; those, like Halley, with orbits lasting less than 200 years. By the 1970s, the rate at which short-period comets were being discovered was becoming increasingly inconsistent with their having emerged solely from the Oort cloud.[22] For an Oort cloud object to become a short-period comet, it would first have to be captured by the giant planets. In 1980, in the Monthly Notices of the Royal Astronomical Society, Julio Fernandez stated that for every short-period comet to be sent into the inner Solar System from the Oort cloud, 600 would have to be ejected into interstellar space. He speculated that a comet belt from between 35 and 50 AU would be required to account for the observed number of comets.[23] Following up on Fernandez's work, in 1988 the Canadian team of Martin Duncan, Tom Quinn and Scott Tremaine ran a number of computer simulations to determine if all observed comets could have arrived from the Oort cloud. They found that the Oort cloud could not account for all short-period comets, particularly as short-period comets are clustered near the plane of the Solar System, whereas Oort-cloud comets tend to arrive from any point in the sky. With a belt as Fernandez described it added to the formulations, the simulations matched observations.[24] Reportedly because the words "Kuiper" and "comet belt" appeared in the opening sentence of Fernandez's paper, Tremaine named this hypothetical region the "Kuiper belt".[25]

## Discovery

The array of telescopes atop Mauna Kea, with which the Kuiper belt was discovered

In 1987, astronomer David Jewitt, then at MIT, became increasingly puzzled by "the apparent emptiness of the outer Solar System".[6] He encouraged then-graduate student Jane Luu to aid him in his endeavour to locate another object beyond Pluto's orbit, because, as he told her, "If we don't, nobody will."[26] Using telescopes at the Kitt Peak National Observatory in Arizona and the Cerro Tololo Inter-American Observatory in Chile, Jewitt and Luu conducted their search in much the same way as Clyde Tombaugh and Charles Kowal had, with a blink comparator.[26] Initially, examination of each pair of plates took about eight hours,[27] but the process was sped up with the arrival of electronic charge-coupled devices or CCDs, which, though their field of view was narrower, were not only more efficient at collecting light (they retained 90 percent of the light that hit them, rather than the ten percent achieved by photographs) but allowed the blinking process to be done virtually, on a computer screen. Today, CCDs form the basis for most astronomical detectors.[28] In 1988, Jewitt moved to the Institute of Astronomy at the University of Hawaii. Luu later joined him to work at the University of Hawaii's 2.24 m telescope at Mauna Kea.[29] Eventually, the field of view for CCDs had increased to 1024 by 1024 pixels, which allowed searches to be conducted far more rapidly.[30] Finally, after five years of searching, on August 30, 1992, Jewitt and Luu announced the "Discovery of the candidate Kuiper belt object" (15760) 1992 $QB_1$.[6] Six months later, they discovered a second object in the region, (181708) 1993 FW.[31]

Studies since the trans-Neptunian region was first charted have shown that the region now called the Kuiper belt is not the point of origin for short-period comets, but that they instead derive from a linked population called the scattered disc. The scattered disc was created when Neptune migrated outward into the proto-Kuiper belt, which at the time was much closer to the Sun, and left in its wake a population of dynamically stable objects which could never be affected by its orbit (the Kuiper belt proper), and a population whose perihelia are close enough that Neptune can still disturb them as it travels around the Sun (the scattered disc). Because the scattered disc is dynamically active and the Kuiper belt relatively dynamically stable, the scattered disc is now seen as the most likely point of origin for periodic comets.[8]

## Name

Astronomers sometimes use the alternative name Edgeworth–Kuiper belt to credit Edgeworth, and KBOs are occasionally referred to as EKOs. However, Brian Marsden claims that neither deserves true credit: "Neither Edgeworth nor Kuiper wrote about anything remotely like what we are now seeing, but Fred Whipple did."[32] Conversely, David Jewitt comments that, "If anything ... Fernandez most nearly deserves the credit for predicting the Kuiper Belt."[13]

KBOs are sometimes called **kuiperoids**, a name suggested by Clyde Tombaugh.[33] The term **trans-Neptunian object** (TNO) is recommended for objects in the belt by several scientific groups because the term is less controversial than all others—it is not a synonym though, as TNOs include all objects orbiting the Sun past the orbit of Neptune, not just those in the Kuiper belt.

# Origins

The precise origins of the Kuiper belt and its complex structure are still unclear, and astronomers are awaiting the completion of several wide-field survey telescopes such as Pan-STARRS and the future LSST, which should reveal many currently unknown KBOs. These surveys will provide data that will help determine answers to these questions.[2]

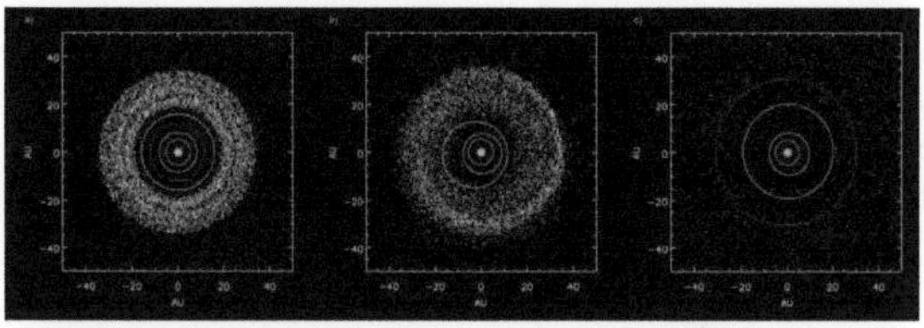

Simulation showing outer planets and Kuiper belt: a) before Jupiter/Saturn 2:1 resonance, b) scattering of Kuiper belt objects into the Solar System after the orbital shift of Neptune, c) after ejection of Kuiper belt bodies by Jupiter

The Kuiper belt is believed to consist of planetesimals; fragments from the original protoplanetary disc around the Sun that failed to fully coalesce into planets and instead formed into smaller bodies, the largest less than 3000 kilometres ( mi) in diameter.

Modern computer simulations show the Kuiper belt to have been strongly influenced by Jupiter and Neptune, and also suggest that neither Uranus nor Neptune could have formed in their present positions, as too little primordial matter existed at that range to produce objects of such high mass. Instead, these planets are believed to have formed closer to Jupiter. Scattering of planetesimals early in the Solar System's history would have led to migration of the orbits of the giant planets; Saturn, Uranus and Neptune drifted outwards while Jupiter drifted inwards. Eventually, the orbits shifted to the point where Jupiter and Saturn reached an exact 2:1 resonance; Jupiter orbited the Sun twice for every one Saturn orbit. The gravitational repercussions of such a resonance ultimately disrupted the orbits of Uranus and Neptune, causing Neptune's orbit to become more eccentric and move outward into the primordial planetesimal disk, which sent the disk into temporary chaos.[34] [35] [36] As Neptune's orbit expanded, it excited and scattered many TNO planetesimals into higher and more eccentric orbits.[37] Many more were scattered inward, often to be scattered again and in some cases ejected by Jupiter. The process is thought to have reduced the primordial Kuiper belt population by 99% or more, and to have shifted the distribution of the surviving members outward.[36]

However, this currently most popular model, the "Nice model", still fails to account for some of the characteristics of the distribution and, quoting one of the scientific articles,[38] the problems "continue to challenge analytical techniques and the fastest numerical modeling hardware and software". The model predicts a higher average eccentricity in classical KBO orbits than is observed (0.10–0.13 versus 0.07).[36] The frequency of paired objects, many of which are far apart and loosely bound, also poses a problem for the model.[39]

## Structure

At its fullest extent, including its outlying regions, the Kuiper belt stretches from roughly 30 to 55 AU. However, the main body of the belt is generally accepted to extend from the 2:3 resonance (see below) at 39.5 AU to the 1:2 resonance at roughly 48 AU.[40] The Kuiper belt is quite thick, with the main concentration extending as much as ten degrees outside the ecliptic plane and a more diffuse distribution of objects extending several times farther. Overall it more resembles a torus or doughnut than a belt.[41] Its mean position is inclined to the ecliptic by 1.86 degrees.[42]

Dust in the Kuiper belt creates a faint infrared disk.

The presence of Neptune has a profound effect on the Kuiper belt's structure due to orbital resonances. Over a timescale comparable to the age of the Solar System, Neptune's gravity destabilises the orbits of any objects which happen to lie in certain regions, and either sends them into the inner Solar System or out into the scattered disc or interstellar space. This causes the Kuiper belt to possess pronounced gaps in its current layout, similar to the Kirkwood gaps in the asteroid belt. In the region between 40 and 42 AU, for instance, no objects can retain a stable orbit over such times, and any observed in that region must have migrated there relatively recently.[43]

### Classical belt

Between the 2:3 and 1:2 resonances with Neptune, at approximately 42–48 AU, the gravitational influence of Neptune is negligible, and objects can exist with their orbits essentially unmolested. This region is known as the classical Kuiper belt, and its members comprise roughly two thirds of KBOs observed to date.[44] [45] Because the first modern KBO discovered, (15760) 1992 $QB_1$, is considered the prototype of this group, classical KBOs are often referred to as cubewanos ("Q-B-1-os").[46] [47] The guidelines established by the IAU demand that classical KBOs be given names of mythological beings associated with creation.[48]

The classical Kuiper belt appears to be a composite of two separate populations. The first, known as the "dynamically cold" population, has orbits much like the planets; nearly circular, with an orbital eccentricity of less than 0.1, and with relatively low inclinations up to about 10° (they lie close to the plane of the Solar System rather than at an angle). The second, the "dynamically hot" population, has orbits much more inclined to the ecliptic, by up to 30°. The two populations have been named this way not because of any major difference in temperature, but from analogy to particles in a gas, which increase their relative velocity as they become heated up.[49] The two populations not only possess different orbits, but different colors; the cold population is markedly redder than the hot. If this is a reflection of different compositions, it suggests they formed in different regions. The hot population is believed to have formed near Jupiter, and to have been ejected out by movements among the gas giants. The cold population, on the other hand, has been proposed to have formed more or less in its current position, although it might also have been later swept outwards by Neptune during its migration,[2] [50] particularly if Neptune's eccentricity was transiently increased.[36] While the Nice model appears to be able to at least partially explain a compositional difference, it has also been suggested the color difference may reflect differences in surface evolution.[36]

## Resonances

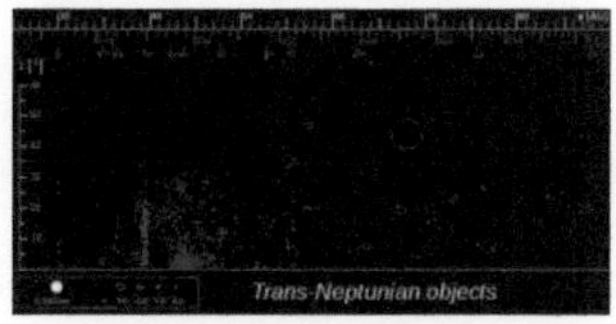


Distribution of cubewanos (blue), Resonant trans-Neptunian objects (red) and near scattered objects (grey).

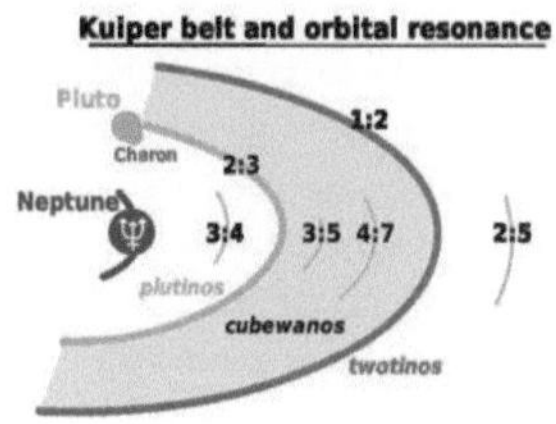


Orbit classification (schematic of semi-major axes)

When an object's orbital period is an exact ratio of Neptune's (a situation called a mean motion resonance), then it can become locked in a synchronised motion with Neptune and avoid being perturbed away if their relative alignments are appropriate. If, for instance, an object is in just the right kind of orbit so that it orbits the Sun two times for every three Neptune orbits, and if it reaches perihelion with Neptune a quarter of an orbit away from it, then whenever it returns to perihelion, Neptune will always be in about the same relative position as it began, since it will have completed 1½ orbits in the same time. This is known as the 2:3 (or 3:2) resonance, and it corresponds to a characteristic semi-major axis of about 39.4 AU. This 2:3 resonance is populated by about 200 known objects,[51] including Pluto together with its moons. In recognition of this, the members of this family are known as plutinos. Many plutinos, including Pluto, have orbits which cross that of Neptune, though their resonance means they can never collide. Plutinos have high orbital eccentricities, suggesting that they are not native to their current positions but were instead thrown haphazardly into their orbits by the migrating Neptune.[52] IAU guidelines dictate that all plutinos must, like Pluto, be named for underworld deities.[48] The 1:2 resonance (whose objects complete half an orbit for each of Neptune's) corresponds to semi-major axes of ~47.7AU, and is sparsely populated.[53] Its residents are sometimes referred to as twotinos. Other resonances also exist at 3:4, 3:5, 4:7 and 2:5.[54] Neptune possesses a number of trojan objects, which occupy its $L_4$ and $L_5$ points; gravitationally stable regions leading and trailing it in its orbit. Neptune trojans are often described as being in a 1:1 resonance with Neptune. Neptune trojans typically have very stable orbits.

Additionally, there is a relative absence of objects with semi-major axes below 39 AU which cannot apparently be explained by the present resonances. The currently accepted hypothesis for the cause of this is that as Neptune migrated outward, unstable orbital resonances moved gradually through this region, and thus any objects within it were swept up, or gravitationally ejected from it.[55]

## "Kuiper cliff"

The 1:2 resonance appears to be an edge beyond which few objects are known. It is not clear whether it is actually the outer edge of the classical belt or just the beginning of a broad gap. Objects have been detected at the 2:5 resonance at roughly 55 AU, well outside the classical belt; however, predictions of a large number of bodies in classical orbits between these resonances have not been verified through observation.[52]

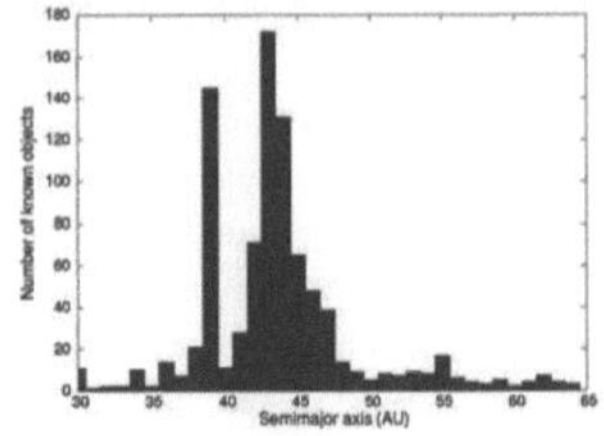


Graph showing the numbers of KBOs for a given distance from the Sun. The plutinos are the "spike" at 39 AU, while the classicals are between 42 and 47 AU, the twotinos are at 48 AU, and the 5:2 resonance is at 55 AU.

Earlier models of the Kuiper belt had suggested that the number of large objects would increase by a factor of two beyond 50 AU,[56] so this sudden drastic falloff, known as the "Kuiper cliff", was completely unexpected, and its cause, to date, is unknown. In 2003, Bernstein and Trilling *et al.* found evidence that the rapid decline in objects of 100 km or more in radius beyond 50 AU is real, and not due to observational bias. Possible explanations include that material at that distance is too scarce or too scattered to accrete into large objects, or that subsequent processes removed or destroyed those which did form.[57] Patryk Lykawka of Kobe University has claimed that the gravitational attraction of an unseen large planetary object, perhaps the size of Earth or Mars, might be responsible.[58] [59]

# Composition

Studies of the Kuiper belt since its discovery have generally indicated that its members are primarily composed of ices: a mixture of light hydrocarbons (such as methane), ammonia, and water ice,[60] a composition they share with comets.[61] The low densities observed in those KBOs whose diameter is known, (less than 1 g $cm^{-3}$) is consistent with an icy makeup.[60] The temperature of the belt is only about 50K,[62] so many compounds that would be gaseous closer to the Sun remain solid.

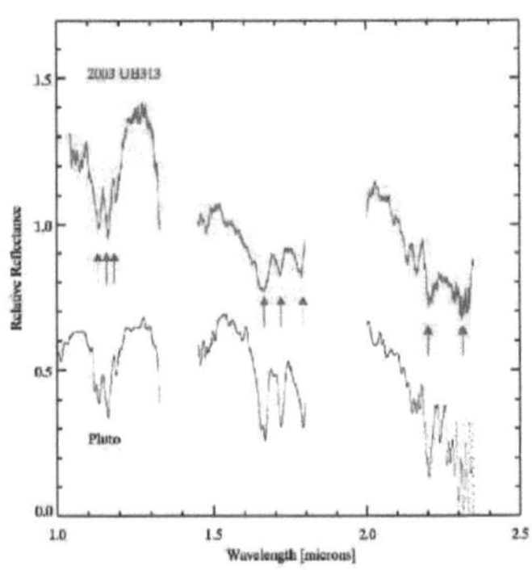


The infrared spectra of both Eris and Pluto, highlighting their common methane absorption lines

Due to their small size and extreme distance from Earth, the chemical makeup of KBOs is very difficult to determine. The principal method by which astronomers determine the composition of a celestial object is spectroscopy. When an object's light is broken into its component colors, an image akin to a rainbow is formed. This image is called a spectrum. Different substances absorb light at different wavelengths, and when the spectrum for a specific object is unravelled, dark lines (called absorption lines) appear where the substances within it have absorbed that particular wavelength of light. Every element or compound has its own unique spectroscopic signature, and by reading an object's full spectral "fingerprint", astronomers can determine what it is made of.

Initially, such detailed analysis of KBOs was impossible, and so astronomers were only able to determine the most basic facts about their makeup, primarily their color.[63] These first data showed a broad range of colors among KBOs, ranging from neutral grey to deep red.[64] This suggested that their surfaces were composed of a wide range of compounds, from dirty ices to hydrocarbons.[64] This diversity was startling, as astronomers had expected KBOs to be uniformly dark, having lost most of their volatile ices to the effects of cosmic rays.[65] Various solutions were suggested for this discrepancy, including resurfacing by impacts or outgassing.[63] However, Jewitt and Luu's spectral analysis of the known Kuiper belt objects in 2001 found that the variation in color was too extreme to be easily explained by random impacts.[66]

Although to date most KBOs still appear spectrally featureless due to their faintness, there have been a number of successes in determining their composition.[62] In 1996, Robert H. Brown *et al.* obtained spectroscopic data on the KBO 1993 SC, revealing its surface composition to be markedly similar to that of Pluto, as well as Neptune's moon Triton, possessing large amounts of methane ice.[67]

Water ice has been detected in several KBOs, including 1996 $TO_{66}$,[68] 38628 Huya and 20000 Varuna.[69] In 2004, Mike Brown *et al.* determined the existence of crystalline water ice and ammonia hydrate on one of the largest known KBOs, 50000 Quaoar. Both of these substances would have been destroyed over the age of the Solar System, suggesting that Quaoar had been recently resurfaced, either by internal tectonic activity or by meteorite impacts.[62]

## Mass and size distribution

Despite its vast extent, the collective mass of the Kuiper belt is relatively low. The total mass is estimated to range between a 25th and 10th the mass of the Earth[70] with some estimates placing it at a thirtieth an Earth mass.[71] Conversely, models of the Solar System's formation predict a collective mass for the Kuiper belt of 30 Earth masses.[2] This missing >99% of the mass can hardly be dismissed, as it is required for the accretion of any KBOs larger than 100 km (62 mi) in diameter. If the Kuiper belt had always had its current low density these large objects simply could not have formed.[2] Moreover, the eccentricity and inclination of current orbits makes the encounters quite "violent" resulting in destruction rather than accretion. It appears that either the current residents of the Kuiper belt have been created closer to the Sun or some mechanism dispersed the original mass. Neptune's current influence is too weak to explain such a massive "vacuuming", though the Nice model proposes that it could have been the cause of mass removal in the past. While the question remains open, the conjectures vary from a passing star scenario to grinding of smaller objects, via collisions, into dust small enough to be affected by solar radiation.[50]

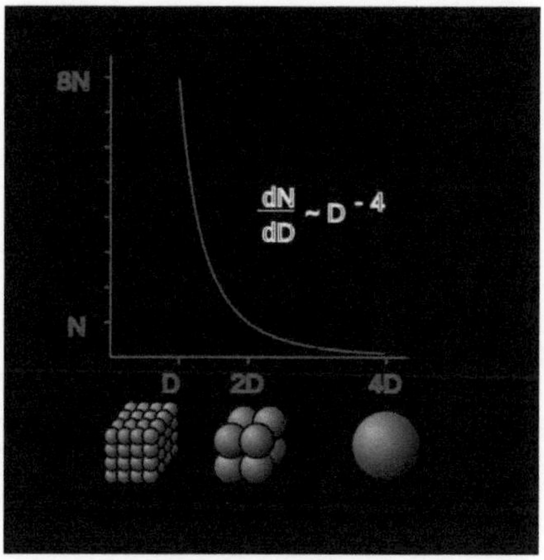

Illustration of the power law.

Bright objects are rare compared with the dominant dim population, as expected from accretion models of origin, given that only some objects of a given size would have grown further. This relationship N(D), the population expressed as a function of the diameter, referred to as brightness slope, has been confirmed by observations. The slope is inversely proportional to some power of the diameter D.

where the current measures[72] give q = 4 ±0.5.

Less formally, there are for instance 8 ($=2^3$) times more objects in 100–200 km range than objects in 200–400 km range. In other words, for every object with the diameter of 1000 km (621 mi) there should be around 1000 ($=10^3$) objects with diameter of 100 km (62 mi).

The law is expressed in this differential form rather than as a cumulative cubic relationship, because only the middle part of the slope can be measured; the law must break at smaller sizes, beyond the current measure.

Of course, only the magnitude is actually known, the size is inferred assuming albedo (not a safe assumption for larger objects).

Since January 2010, the smallest Kuiper belt object discovered to date spans 980 m across.[73]

## Scattered objects

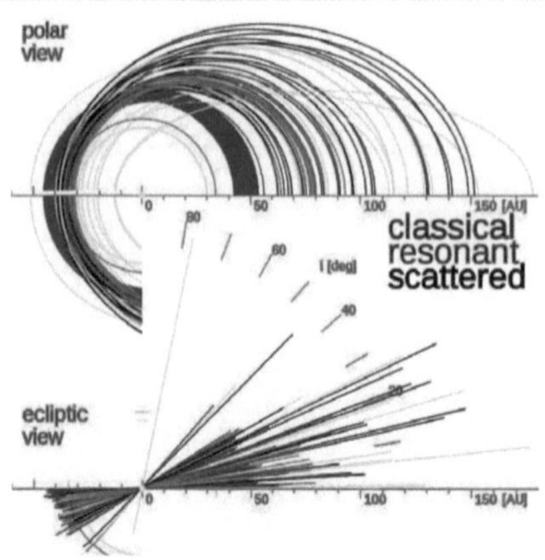


Comparison of the orbits of scattered disc objects (black), classical KBOs (blue), and 2:5 resonant objects (green). Orbits of other KBOs are gray.

The scattered disc is a sparsely populated region, overlapping with the Kuiper belt but extending as far as 100 AU and farther. Scattered disc objects (SDOs) travel in highly elliptical orbits, usually also highly inclined to the ecliptic. Most models of Solar System formation show both KBOs and SDOs first forming in a primordial comet belt, while later gravitational interactions, particularly with Neptune, sent the objects spiraling outward; some into stable orbits (the KBOs) and some into unstable orbits, becoming the scattered disc.[8] Due to its unstable nature, the scattered disc is believed to be the point of origin for many of the Solar System's short-period comets. Their dynamic orbits occasionally force them into the inner Solar System, becoming first centaurs, and then short-period comets.[8]

According to the Minor Planet Center, which officially catalogues all trans-Neptunian objects, a KBO, strictly speaking, is any object that orbits exclusively within the defined Kuiper belt region regardless of origin or composition. Objects found outside the belt are classed as scattered objects.[74] However, in some scientific circles the term "Kuiper belt object" has become synonymous with any icy minor planet native to the outer Solar System believed to have been part of that initial class, even if its orbit during the bulk of Solar System history has been beyond the Kuiper belt (e.g. in the scattered-disc region). They often describe scattered disc objects as "scattered Kuiper belt objects".[75] Eris, which is known to be more massive than Pluto, is often referred to as a KBO, but is technically an SDO.[74] A consensus among astronomers as to the precise definition of the Kuiper belt has yet to be reached, and this issue remains unresolved.

The centaurs, which are not normally considered part of the Kuiper belt, are also believed to be scattered objects, the only difference being that they were scattered inward, rather than outward. The Minor Planet Center groups the centaurs and the SDOs together as scattered objects.[74]

### Triton

Neptune's moon Triton

During its period of migration, Neptune is thought to have captured one of the larger KBOs and set it in orbit around itself. This is its moon Triton, which is the only large moon in the Solar System to have a retrograde orbit; it orbits in the opposite direction to Neptune's rotation. This suggests that, unlike the large moons of Jupiter and Saturn, which are thought to have coalesced from spinning discs of material encircling their young parent planets, Triton was a fully formed body that was captured from surrounding space. Gravitational capture of an object is not easy; it requires some mechanism to slow the object down enough to be snared by the larger object's gravity. Triton may have encountered Neptune as part of a binary (many KBOs are members of binaries; see below); ejection of the other member of the binary by Neptune could then explain Triton's capture.[76] Triton is only slightly larger than Pluto, and spectral analysis of both worlds shows that they are largely composed of similar materials, such as methane and carbon monoxide. All this points to the conclusion that Triton was once a KBO that was captured by Neptune during its outward migration.[77]

## Largest KBOs

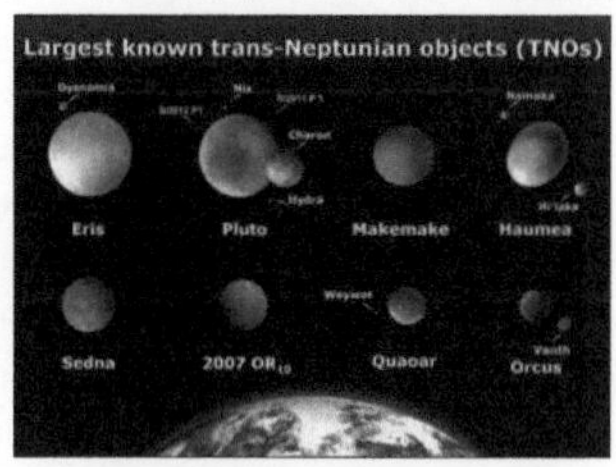


10}}, Quaoar, Orcus, and Earth. These eight trans-Neptunian objects have the brightest absolute magnitudes, although several other TNOs have been found to be physically larger than Orcus, and several more may yet be found.

Since the year 2000, a number of KBOs with diameters of between 500 and 1500 km (932 mi), more than half that of Pluto, have been discovered. 50000 Quaoar, a classical KBO discovered in 2002, is over 1,200 km across. Makemake (originally (136472) 2005 $FY_9$, nicknamed "Easterbunny") and Haumea (originally (136108) 2003 $EL_{61}$, nicknamed "Santa"), both announced on July 29, 2005, are larger still. Other objects, such as 28978 Ixion (discovered in 2001) and 20000 Varuna (discovered in 2000) measure roughly 500 km (311 mi) across.[2]

### Pluto

The discovery of these large KBOs in similar orbits to Pluto led many to conclude that, bar its relative size, Pluto was not particularly different from other members of the Kuiper belt. Not only did these objects approach Pluto in size, but many also possessed satellites, and were of similar composition (methane and carbon monoxide have been found both on Pluto and on the largest KBOs).[2] Thus, just as Ceres was considered a planet before the discovery of its fellow asteroids, some began to suggest that Pluto might also be reclassified.

The issue was brought to a head by the discovery of Eris, an object in the scattered disc far beyond the Kuiper belt, that is now known to be 27 percent more massive than Pluto.[78] In response, the International Astronomical Union (IAU), was forced to define what a planet is for the first time, and in so doing included in their definition that a planet must have "cleared the neighbourhood around its orbit".[79] As Pluto shared its orbit with so many KBOs, it was deemed not to have cleared its orbit, and was thus reclassified from a planet to a member of the Kuiper belt.

Although Pluto is currently the largest KBO, there are two known larger objects currently outside the Kuiper belt which probably originated in the Kuiper belt. These are Eris and Neptune's moon Triton (which, as explained above, is probably a captured KBO).

As of 2008, only five objects in the Solar System, Ceres, Eris, and the KBOs Pluto, Makemake and Haumea, are listed as dwarf planets by the IAU. However, 90482 Orcus, 28978 Ixion and many other Kuiper-belt objects are large enough to be in hydrostatic equilibrium; most of them will probably qualify when more is known about them.[80] [81] [82]

### Satellites

Of the four largest TNOs, three (Eris, Pluto, and Haumea) possess satellites, and two have more than one. A higher percentage of the larger KBOs possess satellites than the smaller objects in the Kuiper belt, suggesting that a different formation mechanism was responsible.[83] There are also a high number of binaries (two objects close enough in mass to be orbiting "each other") in the Kuiper belt. The most notable example is the Pluto–Charon binary, but it is estimated that around 11 percent of KBOs exist in binaries.[84]

## Exploration

Artist's conception of *New Horizons* at Pluto

On January 19, 2006, the first spacecraft mission to explore the Kuiper belt, *New Horizons*, was launched. The mission, headed by Alan Stern of the Southwest Research Institute, will arrive at Pluto on July 14, 2015, and, circumstances permitting, will continue on to study another as-yet undetermined KBO. Any KBO chosen will be between 25 and 55 miles (40 to 90 km) in diameter and, ideally, white or grey, to contrast with Pluto's reddish color.[85] John Spencer, an astronomer on the *New Horizons* mission team, says that no target for a post-Pluto Kuiper belt encounter has yet been selected, as they are awaiting data from the Pan-STARRS survey project to ensure as wide a field of options as possible.[86] The Pan-STARRS project, partially operational since May 2010,[87] will, when fully online, survey the entire sky with four 1.4 gigapixel digital cameras to detect any moving objects, from near-earth objects to KBOs.[88] To speed up the detection process, the New Horizons team established Ice Hunters, a citizen science project that allowed members of the public to participate in the search for suitable KBO targets;[89] [90] [91] the project has subsequently been transferred to another site, Ice Investigators,[92] produced by CosmoQuest.[93]

## Other Kuiper belts

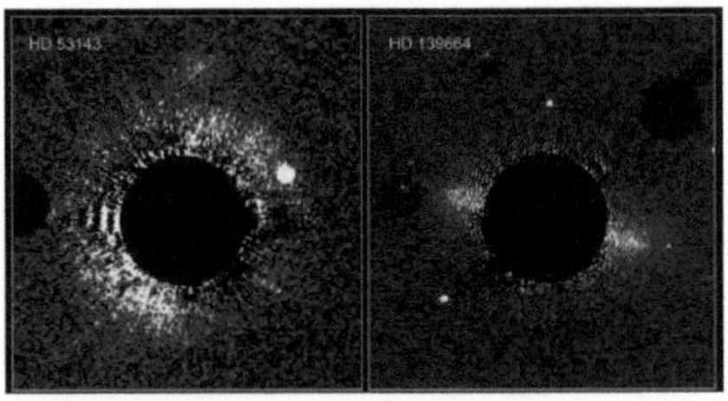

The debris disks around the stars HD 139664 and HD 53143. The black central circle is produced by the camera's coronagraph, which hides the central star to allow the much fainter disks to be seen.

By 2006, astronomers had resolved dust disks believed to be Kuiper belt-like structures around nine stars other than the Sun. They appear to fall into two categories: wide belts, with radii of over 50 AU, and narrow belts (like our own Kuiper belt) with radii of between 20 and 30 AU and relatively sharp boundaries.[94] Beyond this, 15–20% of solar-type stars have an observed infrared excess which is believed to indicate massive Kuiper belt-like structures.[95] Most known debris discs around other stars are fairly young, but the two images on the right, taken by the Hubble Space Telescope in January 2006, are old enough (roughly 300 million years) to have settled into stable configurations. The left image is a "top view" of a wide belt, and the right image is an "edge view" of a narrow belt.[94] [96] Supercomputer simulations of dust in the Kuiper belt suggest that when it was younger, it may have resembled the narrow rings seen around younger stars.[97]

## See also

- List of possible dwarf planets
- List of trans-Neptunian objects

## Notes

The literature is inconsistent in the usage of the terms "scattered disc" and "Kuiper belt". For some, they are distinct populations; for others, the scattered disc is part of the Kuiper belt. Authors may even switch between these two uses in a single publication.[98] Because the International Astronomical Union's Minor Planet Center, the body responsible for cataloguing minor planets in the Solar System, makes the distinction,[99] the current editorial choice for Wikipedia articles on the trans-Neptunian region is to make this distinction as well. This choice means that, on Wikipedia, Eris, the largest known trans-Neptunian object, is not part of the Kuiper belt, and so Pluto becomes the largest Kuiper belt object.

## References


[1] Alan Stern; Colwell, Joshua E. (1997). "Collisional Erosion in the Primordial Edgeworth-Kuiper Belt and the Generation of the 30–50 AU Kuiper Gap". *The Astrophysical Journal* **490** (2): 879–882. Bibcode 1997ApJ...490..879S. doi:10.1086/304912.

[2] Audrey Delsanti and David Jewitt. "The Solar System Beyond The Planets" (http://web.archive.org/web/20070925203400/http://www.ifa.hawaii.edu/faculty/jewitt/papers/2006/DJ06.pdf). *Institute for Astronomy, University of Hawaii*. Archived from the original (http://www2.ess.ucla.edu/~jewitt/papers/2006/DJ06.pdf) on September 25, 2007. . Retrieved March 9, 2007.

[3] Krasinsky, G. A.; Pitjeva, E. V.; Vasilyev, M. V.; Yagudina, E. I. (July 2002). "Hidden Mass in the Asteroid Belt". *Icarus* **158** (1): 98–105. Bibcode 2002Icar..158...98K. doi:10.1006/icar.2002.6837.

[4] Johnson, Torrence V.; and Lunine, Jonathan I.; *Saturn's moon Phoebe as a captured body from the outer Solar System*, Nature, Vol. 435, pp. 69–71

[5] Craig B. Agnor & Douglas P. Hamilton (2006). "Neptune's capture of its moon Triton in a binary-planet gravitational encounter" (http://web.archive.org/web/20070621182809/http://www.es.ucsc.edu/~cagnor/papers_pdf/2006AgnorHamilton.pdf). *Nature*. Archived from the original (http://www.es.ucsc.edu/~cagnor/papers_pdf/2006AgnorHamilton.pdf) on June 21, 2007. . Retrieved June 20, 2006.

[6] Jewitt, David; Luu, Jane (1993). "Discovery of the candidate Kuiper belt object 1992 QB1". *Nature* **362** (6422): 730. Bibcode 1993Natur.362..730J. doi:10.1038/362730a0.

[7] NEW HORIZONS *The PI's Perspective* (http://pluto.jhuapl.edu/overview/piPerspective.php?page=piPerspective_08_24_2012)

[8] Harold F. Levison, Luke Donnes (2007). "Comet Populations and Cometary Dynamics". In Lucy Ann Adams McFadden, Paul Robert Weissman, Torrence V. Johnson. *Encyclopedia of the Solar System* (2nd ed.). Amsterdam; Boston: Academic Press. pp. 575–588. ISBN 0-12-088589-1.

[9] Gérard FAURE (2004). "Description of the System of Asteroids as of May 20, 2004" (http://web.archive.org/web/20070529003558/http://www.astrosurf.com/aude/map/us/AstFamilies2004-05-20.htm). Archived from the original (http://www.astrosurf.com/aude/map/us/AstFamilies2004-05-20.htm) on May 29, 2007. . Retrieved June 1, 2007.

[10] "What is improper about the term "Kuiper belt"? (or, Why name a thing after a man who didn't believe its existence?)" (http://www.icq.eps.harvard.edu/kb.html). *International Comet Quarterly*. . Retrieved October 24, 2010.

[11] John Davies (2001). *Beyond Pluto: Exploring the outer limits of the solar system*. Cambridge University Press. xii.

[12] Davies, p. 2

[13] David Jewitt. "WHY "KUIPER" BELT?" (http://www2.ess.ucla.edu/~jewitt/kb/gerard.html). *University of Hawaii*. . Retrieved June 14, 2007.

[14] Davies, p. 14

[15] Rao, M. M. (1964). "Decomposition of Vector Measures" (http://www.pnas.org/cgi/reprint/51/5/711.pdf). *Proceedings of the National Academy of Sciences* **51** (5): 771. Bibcode 1964PNAS...51..771R. doi:10.1073/pnas.51.5.771. .

[16] CT Kowal, W Liller, BG Marsden; Liller; Marsden (1977). "The discovery and orbit of /2060/ Chiron". *Hale Observatories, Harvard–Smithsonian Center for Astrophysics* **81**: 245. Bibcode 1979IAUS...81..245K.

[17] JV Scotti, DL Rabinowitz, CS Shoemaker, EM Shoemaker, DH Levy, TM King, EF Helin, J Alu, K Lawrence, RH McNaught, L Frederick, D Tholen, BEA Mueller; Rabinowitz; Shoemaker; Shoemaker; Levy; King; Helin; Alu et al. (1992). "1992 AD". *IAU Circ.* **5434**: 1. Bibcode 1992IAUC.5434....1S.

[18] Horner, J.; Evans, N.W.; Bailey, M. E. (2004). "Simulations of the Population of Centaurs I: The Bulk Statistics". *MNRAS* **354** (3): 798–810. arXiv:astro-ph/0407400. Bibcode 2004MNRAS.354..798H. doi:10.1111/j.1365-2966.2004.08240.x.

[19] Davies p. 38

[20] David Jewitt (2002). "From Kuiper Belt Object to Cometary Nucleus: The Missing Ultrared Matter". *The Astronomical Journal* **123** (2): 1039–1049. Bibcode 2002AJ....123.1039J. doi:10.1086/338692.

[21] Oort, J. H. (1950). "The structure of the cloud of comets surrounding the Solar System and a hypothesis concerning its origin". *Bull. Astron. Inst. Neth.* **11**: 91. Bibcode 1950BAN....11...91O.

[22] Davies p. 39

[23] JA Fernandez (1980). "On the existence of a comet belt beyond Neptune". *Monthly Notices of the Royal Astronomical Society* **192**: 481. Bibcode 1980MNRAS.192..481F.

[24] M. Duncan, T. Quinn, and S. Tremaine (1988). "The origin of short-period comets". *The Astrophysical Journal* **328**: L69. Bibcode 1988ApJ...328L..69D. doi:10.1086/185162.

[25] Davies p. 191

[26] Davies p. 50

[27] Davies p. 51

[28] Davies pp. 52, 54, 56

[29] Davies pp. 57, 62

[30] Davies p. 65

[31] BS Marsden; Jewitt; Marsden (1993). "1993 FW". *Minor Planet Center* **5730**: 1. Bibcode 1993IAUC.5730....1L.

[32] Davies p. 199

[33] Clyde Tombaugh, "The Last Word", Letters to the Editor, *Sky & Telescope*, December, 1994, p. 8

[34] Hansen, K. (7 June 2005). "Orbital shuffle for early solar system" (http://www.geotimes.org/june05/WebExtra060705.html). *Geotimes*. . Retrieved 2007-08-26.

[35] Tsiganis, K.; Gomes, R.; Morbidelli, A.; Levison, H. F. (2005). "Origin of the orbital architecture of the giant planets of the Solar System". *Nature* **435** (7041): 459–461. Bibcode 2005Natur.435..459T. doi:10.1038/nature03539. PMID 15917800.

[36] Levison, H. F.; Morbidelli, A.; Van Laerhoven, C.; Gomes, R. (2008). "Origin of the structure of the Kuiper belt during a dynamical instability in the orbits of Uranus and Neptune". *Icarus* **196** (1): 258–273. arXiv:0712.0553. Bibcode 2008Icar..196..258L. doi:10.1016/j.icarus.2007.11.035.

[37] Thommes, E. W.; Duncan, M. J.; Levison, H. F. (2002). "The Formation of Uranus and Neptune among Jupiter and Saturn". *The Astronomical Journal* **123** (5): 2862. arXiv:astro-ph/0111290. Bibcode 2002AJ....123.2862T. doi:10.1086/339975.

[38] Malhotra, R. (1994). "Nonlinear Resonances in the Solar System". *Physica D* **77**: 289. arXiv:chao-dyn/9406004. Bibcode 1994PhyD...77..289M. doi:10.1016/0167-2789(94)90141-4.

[39] Lovett, R. (2010). "Kuiper Belt may be born of collisions". *Nature*. doi:10.1038/news.2010.522.

[40] M. C. De Sanctis, M. T. Capria, and A. Coradini (2001). "Thermal Evolution and Differentiation of Edgeworth-Kuiper Belt Objects". *The Astronomical Journal* **121** (5): 2792–2799. Bibcode 2001AJ....121.2792D. doi:10.1086/320385.

[41] "Discovering the Edge of the Solar System" (http://web.archive.org/web/20080305191925/http://www.americanscientist.org/template/AssetDetail/assetid/25723/page/2;jsessionid=aaa5LVF0). *American Scientists.org*. 2003. Archived from the original (http://www.americanscientist.org/template/AssetDetail/assetid/25723/page/2;jsessionid=aaa5LVF0) on March 5, 2008. . Retrieved June 23, 2007.

[42] Michael E. Brown, Margaret Pan (2004). "The Plane of the Kuiper Belt". *The Astronomical Journal* **127** (4): 2418–2423. Bibcode 2004AJ....127.2418B. doi:10.1086/382515.

[43] Jean-Marc Petit, Alessandro Morbidelli, Giovanni B. Valsecchi (1998). "Large Scattered Planetesimals and the Excitation of the Small Body Belts" (http://www.obs-nice.fr/morby/papers/6166a.pdf). . Retrieved June 23, 2007.

[44] Lunine, J. (2003). "The Kuiper Belt" (http://www.gsmt.noao.edu/gsmt_swg/SWG_Apr03/The_Kuiper_Belt.pdf). . Retrieved 2007-06-23.

[45] Jewitt, D. (February 2000). "Classical Kuiper Belt Objects (CKBOs)" (http://web.archive.org/web/20070609094740/http://www.ifa.hawaii.edu/~jewitt/kb/kb-classical.html). Archived from the original (http://www2.ess.ucla.edu/~jewitt/kb/kb-classical.html) on 2007-06-09. . Retrieved 2007-06-23.

[46] Murdin, P. (2000). "Cubewano". *The Encyclopedia of Astronomy and Astrophysics*. Bibcode 2000eaa..bookE5403.. doi:10.1888/0333750888/5403. ISBN 0-333-75088-8.

[47] Elliot, J. L.; *et al.* (2005). "The Deep Ecliptic Survey: A Search for Kuiper Belt Objects and Centaurs. II. Dynamical Classification, the Kuiper Belt Plane, and the Core Population" (http://occult.mit.edu/_assets/documents/publications/Elliot2005AJ129.1117.pdf). *The Astronomical Journal* **129**: 1117–1162. Bibcode 2005AJ....129.1117E. doi:10.1086/427395. .

[48] "Naming of Astronomical Objects: Minor Planets" (http://www.iau.org/public_press/themes/naming/#minorplanets). International Astronomical Union. . Retrieved 2008-11-17.

[49] Levison, H. F; Morbidelli, A. (2003). "The formation of the Kuiper belt by the outward transport of bodies during Neptune's migration". *Nature* **426** (6965): 419–421. doi:10.1038/nature02120. PMID 14647375.

[50] Morbidelli, A. (2005). "Origin and Dynamical Evolution of Comets and their Reservoirs". arXiv:astro-ph/0512256 [astro-ph].

[51] "List Of Transneptunian Objects" (http://www.minorplanetcenter.org/iau/lists/TNOs.html). *Minor Planet Center*. . Retrieved June 23, 2007.

[52] Chiang et al. (2003). "Resonance Occupation in the Kuiper Belt: Case Examples of the 5:2 and Trojan Resonances". *The Astronomical Journal* **126** (1): 430–443. arXiv:astro-ph/0301458. Bibcode 2003AJ....126..430C. doi:10.1086/375207.

[53] Wm. Robert Johnston (2007). "Trans-Neptunian Objects" (http://www.johnstonsarchive.net/astro/tnos.html). . Retrieved June 23, 2007.

[54] Davies p. 104

[55] Davies p. 107
[56] E. I. Chiang and M. E. Brown (1999). "Keck Pencil-Beam Survey For Faint Kuiper Belt Objects" (http://www.gps.caltech.edu/~mbrown/papers/ps/kbodeep.pdf). . Retrieved July 1, 2007.
[57] G.M. Bernstein, D.E. Trilling, R.L. Allen, M.E. Brown, M. Holman and R. Malhotra (2004). "The Size Distribution of Trans-Neptunian Bodies" (http://www.gps.caltech.edu/~mbrown/papers/ps/bernstein.pdf). *The Astrophysical Journal* **128** (3): 1364. arXiv:astro-ph/0308467. Bibcode 2004AJ....128.1364B. doi:10.1086/422919. .
[58] Michael Brooks (2007). "13 Things that do not make sense" (http://space.newscientist.com/article.ns?id=mg18524911.600). *NewScientistSpace.com*. . Retrieved June 23, 2007.
[59] Govert Schilling (2008). "The mystery of Planet X" (http://space.newscientist.com/article/mg19726381.600-the-mystery-of-planet-x.html). *New Scientist*. . Retrieved February 8, 2008.
[60] Stephen C. Tegler (2007). "Kuiper Belt Objects: Physical Studies". In Lucy-Ann McFadden *et al.*. *Encyclopedia of the Solar System*. pp. 605–620.
[61] Altwegg, K.; Balsiger, H.; Geiss, J. (1999). *Space Science Reviews* **90**: 3. Bibcode 1999SSRv...90....3A. doi:10.1023/A:1005256607402.
[62] David C. Jewitt & Jane Luu (2004). "Crystalline water ice on the Kuiper belt object (50000) Quaoar" (http://web.archive.org/web/20070621182808/http://www.ifa.hawaii.edu/~jewitt/papers/50000/Quaoar.pdf). Archived from the original (http://www2.ess.ucla.edu/~jewitt/papers/50000/Quaoar.pdf) on June 21, 2007. . Retrieved June 21, 2007.
[63] Dave Jewitt (2004). "Surfaces of Kuiper Belt Objects" (http://web.archive.org/web/20070609094911/http://www.ifa.hawaii.edu/~jewitt/kb/kb-colors.html). *University of Hawaii*. Archived from the original (http://www2.ess.ucla.edu/~jewitt/kb/kb-colors.html) on June 9, 2007. . Retrieved June 21, 2007.
[64] Jewitt, David; Luu, Jane (1998). "Optical-Infrared Spectral Diversity in the Kuiper Belt". *The Astronomical Journal* **115** (4): 1667. Bibcode 1998AJ....115.1667J. doi:10.1086/300299.
[65] Davies p. 118
[66] Jewitt, David C.; Luu, Jane X. (2001). "Colors and Spectra of Kuiper Belt Objects". *The Astronomical Journal* **122** (4): 2099. arXiv:astro-ph/0107277. Bibcode 2001AJ....122.2099J. doi:10.1086/323304.
[67] Brown, R. H.; Cruikshank, DP; Pendleton, Y; Veeder, GJ (1997). "Surface Composition of Kuiper Belt Object 1993SC". *Science* **276** (5314): 937–9. Bibcode 1997Sci...276..937B. doi:10.1126/science.276.5314.937. PMID 9163038.
[68] Brown, Michael E.; Blake, Geoffrey A.; Kessler, Jacqueline E. (2000). "Near-Infrared Spectroscopy of the Bright Kuiper Belt Object 2000 EB173". *The Astrophysical Journal* **543** (2): L163. Bibcode 2000ApJ...543L.163B. doi:10.1086/317277.
[69] Licandro; Oliva; Di MArtino (2001). "NICS-TNG infrared spectroscopy of trans-neptunian objects 2000 EB173 and 2000 WR106". *Astronomy and Astrophysics* **373** (3): L29. arXiv:astro-ph/0105434. Bibcode 2001A&A...373L..29L. doi:10.1051/0004-6361:20010758.
[70] Gladman, Brett; et al (05/April/2001). "The structure of the Kuiper belt". *Astronomical Journal* **122** (2): 1051–1066. Bibcode 2001AJ....122.1051G. doi:10.1086/322080.
[71] Lorenzo Iorio (2007). "Dynamical determination of the mass of the Kuiper Belt from motions of the inner planets of the Solar system". *Monthly Notices of the Royal Astronomical Society* **375** (4): 1311–1314. Bibcode 2007MNRAS.tmp...24I. doi:10.1111/j.1365-2966.2006.11384.x.
[72] Bernstein, G. M.; Trilling, D. E.; Allen, R. L.; Brown, K. E.; Holman, M.; Malhotra, R. (2004). "The size distribution of transneptunian bodies". *The Astronomical Journal* **128** (3): 1364–1390. arXiv:astro-ph/0308467. Bibcode 2004AJ....128.1364B. doi:10.1086/422919.
[73] l, Eric (January 13, 2010). "Hubble Finds Smallest Kuiper Belt Object" (http://www.meteoritesusa.com/meteorite-photos/hubble-finds-smallest-kuiper-belt-object/). *MeteoritesUSA*. . Retrieved January 16, 2011.
[74] "List Of Centaurs and Scattered-Disk Objects" (http://www.minorplanetcenter.org/iau/lists/Centaurs.html). *IAU: Minor Planet Center*. . Retrieved October 27, 2010.
[75] David Jewitt (2005). "The 1000 km Scale KBOs" (http://www2.ess.ucla.edu/~jewitt/kb/big_kbo.html). *University of Hawaii*. . Retrieved July 16, 2006.
[76] Craig B. Agnor & Douglas P. Hamilton (2006). "Neptune's capture of its moon Triton in a binary-planet gravitational encounter" (http://web.archive.org/web/20070621182809/http://www.es.ucsc.edu/~cagnor/papers_pdf/2006AgnorHamilton.pdf). *Nature*. Archived from the original (http://www.es.ucsc.edu/~cagnor/papers_pdf/2006AgnorHamilton.pdf) on June 21, 2007. . Retrieved October 29, 2007.
[77] DALE P. CRUIKSHANK (2004). *TRITON, PLUTO, CENTAURS, AND TRANS-NEPTUNIAN BODIES* (http://books.google.com/?id=MbmiTd3x1UcC&pg=PA421&dq=.+TRITON,+PLUTO,+CENTAURS,+AND+TRANS-NEPTUNIAN+BODIES). Springer. ISBN 978-1-4020-3362-9. . Retrieved June 23, 2007.
[78] Mike Brown (2007). "Dysnomia, the moon of Eris" (http://www.gps.caltech.edu/~mbrown/planetlila/moon/index.html). *CalTech*. . Retrieved June 14, 2007.
[79] "Resolution B5 and B6" (http://www.iau.org/static/resolutions/Resolution_GA26-5-6.pdf). International Astronomical Union. 2006. .
[80] "Ixion" (http://ixion.eightplanets.net/). *eightplanets.net*. . Retrieved June 23, 2007.
[81] John Stansberry; Will Grundy; Mike Brown; Dale Cruikshank; John Spencer; David Trilling; Jean-Luc Margot (2007). "Physical Properties of Kuiper Belt and Centaur Objects: Constraints from Spitzer Space Telescope". arXiv:astro-ph/0702538 [astro-ph].
[82] "IAU Draft Definition of Planet" (http://www.iau.org/iau0601.424.0.html). *IAU*. 2006. . Retrieved October 26, 2007.
[83] Error: Bad DOI specified!
[84] Agnor, C.B.; Hamilton, D.P. (2006). "Neptune's capture of its moon Triton in a binary-planet gravitational encounter" (http://www.astro.umd.edu/~hamilton/research/reprints/AgHam06.pdf). *Nature* **441** (7090): 192–4. Bibcode 2006Natur.441..192A.

doi:10.1038/nature04792. PMID 16688170. .

[85] "New Horizons mission timeline" (http://pluto.jhuapl.edu/mission/mission_timeline.php). *NASA*. . Retrieved March 19, 2011.

[86] Cal Fussman (2006). "The Man Who Finds Planets" (http://discovermagazine.com/2006/may/cover/article_view?b_start:int=3&-C=). *Discover magazine*. . Retrieved August 13, 2007.

[87] Institute for Astronomy, University of Hawai (2010). "PS1 goes Operational and begins Science Mission, May 2010" (http://pan-starrs.ifa.hawaii.edu/public/project-status/PS1 operational.html). . Retrieved August 30, 2010.

[88] "Pan-Starrs: University of Hawaii" (http://pan-starrs.ifa.hawaii.edu/public/home.html). 2005. . Retrieved August 13, 2007.

[89] "Ice Hunters web site" (http://www.icehunters.org). Zooniverse.Org. . Retrieved July 8, 2011.

[90] "Citizen Scientists: Discover a New Horizons Flyby Target" (http://solarsystem.nasa.gov/news/display.cfm?News_ID=37726). NASA. Jun 21, 2011. . Retrieved August 23, 2011.

[91] Lakdawalla, Emily (June 21, 2011). "The most exciting citizen science project ever (to me, anyway)" (http://planetary.org/blog/article/00003073/). The Planetary Society. . Retrieved August 31, 2011.

[92] "Ice Investigators" (http://cosmoquest.org/iceinvestigators/). *web site*. CosmoQuest. 2012. . Retrieved 2012-05-23.

[93] "Finding Ice" (http://cosmoquest.org/blog/2012/05/finding-ice/). *CosmoQuest web site* (http://cosmoquest.org/). CosmoQuest. 2012-05-20. . Retrieved 2012-05-23.

[94] Kalas, Paul; Graham, James R.; Clampin, Mark C.; Fitzgerald, Michael P. (2006). "First Scattered Light Images of Debris Disks around HD 53143 and HD 139664". *The Astrophysical Journal* **637**: L57. arXiv:astro-ph/0601488. Bibcode 2006ApJ...637L..57K. doi:10.1086/500305.

[95] Trilling, D. E.; Bryden, G.; Beichman, C. A.; Rieke, G. H.; Su, K. Y. L.; Stansberry, J. A.; Blaylock, M.; Stapelfeldt, K. R.; Beeman, J. W.; Haller, E. E. (February 2008). "Debris Disks around Sun-like Stars". *The Astrophysical Journal* **674** (2): 1086–1105. arXiv:0710.5498. Bibcode 2008ApJ...674.1086T. doi:10.1086/525514.

[96] "Dusty Planetary Disks Around Two Nearby Stars Resemble Our Kuiper Belt" (http://hubblesite.org/newscenter/archive/releases/2006/05/image/a). 2006. . Retrieved July 1, 2007.

[97] Kuchner, M. J.; Stark, C. C. (2010). "Collisional Grooming Models of the Kuiper Belt Dust Cloud". *The Astronomical Journal* **140** (4): 1007–1019. arXiv:1008.0904. Bibcode 2010AJ....140.1007K. doi:10.1088/0004-6256/140/4/1007.

[98] Weissman and Johnson, 2007, *Encyclopedia of the solar system*, footnote p. 584

[99] IAU: Minor Planet Center (2011-01-03). "List Of Centaurs and Scattered-Disk Objects" (http://www.minorplanetcenter.org/iau/lists/Centaurs.html). Central Bureau for Astronomical Telegrams, Harvard-Smithsonian Center for Astrophysics. . Retrieved 2011-01-03.

## External links and data sources

- Dave Jewitt's page @ UCLA (http://www2.ess.ucla.edu/~jewitt/kb.html)
  - The belt's name (http://www2.ess.ucla.edu/~jewitt/kb/gerard.html)
- Ice Hunters (http://www.icehunters.org/), a citizen science project to help discover a KBO to serve as a followup *New Horizons* mission target after Pluto
- List of short period comets by family (http://www.physics.ucf.edu/~yfernandez/cometlist.html)
- Kuiper Belt Profile (http://solarsystem.nasa.gov/planets/profile.cfm?Object=KBOs) by NASA's Solar System Exploration (http://solarsystem.nasa.gov)
- The Kuiper Belt Electronic Newsletter (http://www.boulder.swri.edu/ekonews/)
- Wm. Robert Johnston's TNO page (http://www.johnstonsarchive.net/astro/tnos.html)
- Minor Planet Center: Plot of the Outer Solar System (http://www.minorplanetcenter.org/iau/lists/OuterPlot.html), illustrating Kuiper gap
- Website of the International Astronomical Union (http://www.iau.org/) (debating the status of TNOs)
- XXVIth General Assembly 2006 (http://www.astronomy2006.com)
- nature.com article: diagram displaying inner solar system, Kuiper Belt, and Oort Cloud (http://www.nature.com/nature/journal/v424/n6949/fig_tab/nature01725_F1.html), taken from Alan Stern, S. (2003). "The evolution of comets in the Oort cloud and Kuiper belt". *Nature* **424** (6949): 639–42. doi:10.1038/nature01725. PMID 12904784.
- SPACE.com: Discovery Hints at a Quadrillion Space Rocks Beyond Neptune (http://www.space.com/scienceastronomy/060814_tno_found.html) (Sara Goudarzi) August 15, 2006 06:13 am ET
- The Outer Solar System (http://www.astronomycast.com/astronomy/episode-64-pluto-and-the-icy-outer-solar-system/) Astronomy Cast episode No. 64, includes full transcript.
- The Kuiper belt (http://365daysofastronomy.org/2009/12/08/december-8th-what-is-the-kuiper-belt/) at 365daysofastronomy.org

# Article Sources and Contributors

**12446_Juliabryant** *Source*: http://en.wikipedia.org/w/index.php?title=12446_Juliabryant *Contributors*: Anypodetos, Avicennasis, Captain panda, Merovingian, Rich Farmbrough, T@nn

**Asteroid_belt** *Source*: http://en.wikipedia.org/w/index.php?title=Asteroid_belt *Contributors*: 1981willy, 21655, 2T, A-giau, Abbalkea2010, Ace45954, Adam Keller, Aesopos, Ahoerstemeier, Aiken drum, Airplaneman, Akshay Lattimardi, Alansohn, AlexiusHoratius, Alfio, Alsandro, Amara, Anarchist42, Andrewlp1991, Antandrus, Apeman, Aponar Kestrel, Arsia Mons, Art LaPella, Arz1969, Avenue, Avicennasis, AxelBoldt, Baa, Bender235, Bewildebeast, Bfissa, Bigger digger, Blacx0, BlastOButter42, Bobymeltics, Bongwarrior, Bryan Derksen, Burzmali, CanadianLinuxUser, Casliber, CharlesC, Cheat2win, Chesnok, ChristTrekker, ChristopherWillis, CielProfond, CirrusHead, Cjc13, Ckatz, Closedmouth, Coalisnotcools, Colonies Chris, Crash Underride, Crazymonkey1123, DABADABADAI, Dabomb87, Danim, Davewild, Dawnseeker2000, Delrayva, Denisarona, Deon, Deor, Deuar, Discospinster, Donarreiskoffer, Dougano, DrKiernan, Dsparks000, Dspitzle, Dune010, Dwcarless, EarnestyEternity, Eccles999174, Ed Poor, Editore99, Edward130603, Egil, El C, El Cid, Emurphy42, Enviroboy, Epbr123, Everyking, Excirial, Extra999, FF2010, Falcon8765, Fcn, Feens, Felyza, Femto, Fjörgynn, Flyguy649, Folajimi, Fotaun, FrankCarroll, Frenchgizmo, Furrykef, Gaius Cornelius, Gap9551, Gareth Griffith-Jones, Garion96, Geeoharee, Gene Nygaard, Geoff Plourde, George100, Gilderien, Gilliam, Gogo Dodo, Goudzovski, Grahamec, Gtg204y, Gurch, GuyQuest, Gwillhickers, Hadseys, Hairy Dude, HalfShadow, Hamiltonstone, Harry, Hashar, HatriX, Hda3ku, Hdt83, Headbomb, HenryLi, Herostratus, Hibernian, Hiberniantears, Hike395, Hmains, Hope(N Forever), Hunnjazal, I like trains beetches, IVAN3MAN, IanOsgood, Ianare, Icairns, Ilikebakedziti, Improve, Incnis Mrsi, Iridescent, J.delanoy, JForget, JHunterJ, JR00576, Jacj, Jackfork, JamesHoadley, Jaredroberts, Jcronen1, Jeff G., Jeremy Visser, Jim1138, Jivecat, Jnestorius, JocK, Joepearson, John Reaves, John of Reading, JorisvS, Jusdafax, Jusjih, Jwh, Ke6jjj, Kejo13, Kheider, Kintetsubuffalo, Kobrabones, Kristaga, Kuru, Kwamikagami, Kyng, Lakers, Lampman, Lanthanum-138, Latulla, Lavateraguy, Lawrie88f, Leor klier, Lhopitalified, Liamdaly620, Lightmouse, Lilly Inhappy, LonelyMarble, Lotu, Luk, Luokehao, MJ94, Malinadams, Manisero399, Marcus Qwertyus, Marek69, Marksurge, Marskell, Martinwilke1980, Materialscientist, Matthewedwards, Maxamegalon2000, McGeddon, Meelar, Megalodon99, Melara..., Merovingian, Metafax1, Mgiganteus1, Michael Devore, Michael Hardy, Michaelas10, Mike s, Milonica, Moneo, Moocha, Morgan Leigh, MrRadioGuy, Mulad, Murgh, Muriel Gottrop, Murlough23, Mxn, Myanw, Myrrhlin, Narayanese, Nephelin, NerdyScienceDude, Nick C, Nick123, Nivix, NorwegianBlue, Numbers123---, Oldmanjenkins93, Oni Ookami Alfador, Ontyx, Oranjblud, Orii, Ost316, P. S. F. Freitas, P30Carl, Palica, Papa November, Paul H., Pembers, Peyre, Pharos, Philip Trueman, Piano non troppo, Pie4all88, Pinethicket, Pizza Puzzle, Pmanderson, PocketDocket, Poppy, Potatoswatter, PowerCS, Praefectorian, Qu Hongcheng, Quackers074, Quendus, Qurq, RJHall, RP88, RSStockdale, RadicalOne, Ramaksoud2000, Ran, Randomblue, Razorflame, Reaper Eternal, Res2216firestar, Retodon8, Revolución, RexNL, Reywas92, Rhowryn, Rich Farmbrough, Richhoncho, Rjwilmsi, Rmhermen, RobertG, Rogermw, Ronhjones, Rorschach, RossPatterson, Rothorpe, Rst20xx, Ruby.red.roses, Ruslik0, Rwflammang, S Whistler, SERK12, SMSshark, Sardanaphalus, Saros136, Sbig10, SchuminWeb, Scog, ScottyBerg, ScottyBoy900Q, Sdornan, Seaphoto, Sengkang, Serendipodous, Shadowcheets, Shadowjams, Shirik, Shshshsh, Siafu, Sjock, Skizzik, Skysmith, Smartech, SnappingTurtle, Snowolf, Some jerk on the Internet, Something14, Springnuts, Squirepants101, Sry85, Stephen e nelson, Subash.chandran007, Sugarfish, Sumone10154, SuperHamster, SuperSpy00bob, TRUTH316, Taed, Tbhotch, Tcncv, Tearlach, Tedickey, Template namespace initialisation script, Teog, TerraFrost, The High Fin Sperm Whale, The Rambling Man, The Tom, Thehelpfulone, Tide rolls, Tiptoety, Titus III, Tombombadil, Tpbradbury, TravisTX, Treisijs, TrevorX, Tuckerresearch, Ufviper, Uncle Dick, Urhixidur, Vanished User 0001, Viriditas, Visor, Vittsadaf, Vsmith, WRK, Waxcaptain45, Wayne Hardman, Wayne Slam, Weedwhacker128, West-J, Wetman, Whatcanbrowndo, WhiteDragon, Wiki alf, Wiki libs, Wikih101, Wikisux, Wiklol, William Avery, Wimt, Wipsenade, Wolfkeeper, WolfmanSF, Worldtraveller, Wtmitchell, X201, Xoxojoccy, Xxgurlxx, Yosef1987, Zachary, Zamphuor, Zanejune1, Zidane tribal, Zé da Silva, $_0$x, 677 anonymous edits

**Asteroid** *Source*: http://en.wikipedia.org/w/index.php?title=Asteroid *Contributors*: (, -- April, 10metreh, 129.128.164.xxx, 129.128.90.xxx, 15lsoucy, 1981willy, 1984, 1wonjae, 23skidoo, 28421u2232nfenfcenc, 2help, 3morand, 95j, A-giau, A4, AATTW, ABF, ACSE, AVand, Abc518, Academic Challenger, Acalamari, Access Denied, Accurizer, Adam Keller, Adambro, Addps4cat, AdjustShift, Adnergje, AgadaUrbanit, Ageekgal, Aggie747, Ahoerstemeier, Ajitem49, Ajraddatz, Alan R. Fisher, Alansohn, Ale jrb, Aleenf1, Alex.muller, Alexf, AlexiusHoratius, Alfio, Allstarecho, AllyUnion, Alsandro, Amaury, Amazins490, Amire80, Anaxial, Andonic, Andre Engels, Andrew Gray, AndrewBuck, Andy4789, AngelOfSadness, Angela, AnimAlu, AnkhMorpork, Ann Stouter, Anna512, Antandrus, Antoni Barau, Anville, Apparition11, Apple K, Arcman, Arjun01, Arkuat, Aster2, Asteroidz R not planetz, Astrobiologist, Ato, Attackpig, Avicennasis, Ayelie, Aznrules, Backslash Forwardslash, Badgernet, Banzai314, Bathanyucolkol, Beemer69, Beland, Ben-Zin, Ben.davis35, Benhocking, Bercziszani, Bevo, Bihco, BilCat, Bill37212, Billymac00, Blanchardb, Bluemask, Bob f it, Bobblewik, Bobizuruncle, Bobo192, Bogey97, Bokac1984, Bomac, Bongwarrior, Brandon, Brion VIBBER, Brockert, Bryan Derksen, Bsimmons666, CBDunkerson, CWenger, CWii, Callanecc, Callumw1, Calvin 1998, Can't sleep, clown will eat me, CapitalR, Captain Yesterday, Captain-tucker, Captainbeefart, Carinemily, Carmichael, Carson69, Casmith 789, Catapult, Catttttt2, Celarnor, Ceranthor, Charles Matthews, Charliesface, CharlotteWebb, Chase me ladies, I'm the Cavalry, Chesnok, ChiZeroOne, Chris goulet, Chuck02, Chunkylefunga, Ck lostsword, Ckatz, Ckruschke, Clovis Sangrail, Cmapm, CommonsDelinker, Conversion script, CoolbeansofGR, Coolkid1001, Courcelles, Craigboy, Creidieki, Cryptic, Css, Curps, CzarB, DARTH SIDIOUS 2, DMacks, DVD R W, DW declare, DaL33T, Daddylonglegs420, Dam sens, Danakil, Daniel, DanielCD, Danim, DanwWiki, Daranak, Dartoe, David0811, DavidLevinson, Davidiad, DazBoz, Dcoetzee, Deagle AP, Dekaels, Dendodge, Deor, Deuar, Dffgd, Diderot, Dina, Discospinster, Divebomb, Dman159753, DocWatson42, Dr. Blofeld, DrGatsby1962, Dragon of the Pants, Drbogdan, Dreish, Drift chambers, Drmies, Dstrike224, Duoduoduo, Dwadejames03, Dysepsion, E. Fokker, ERcheck, ESkog, Editor at Large, Eeekster, Eggoeater, Ejosse1, El C, El Shaday, Eleassar, Electron9, Emerson7, Emijrp, Emote, Emurphy42, Enemyofthestate, EoGuy, Epbr123, Eric-Wester, Ericann726, Erik9, Etacar11, EvenGreenerFish, Evil Monkey, Evil saltine, Evvyleeds, Excirial, Eyrian, FM, Fabiform, Fabiob, Faker12345, Falcon8765, Fartherred, Femto, Ferbtastic, Ffbond, Finlay McWalter, Finngall, Firsfron, Fisherama1, Fld300b, Flex Flint, Florian Blaschke, Flyguy649, FlyingToaster, Folajimi, Foobaz, Fosnez, Fountains of Bryn Mawr, Fourohfour, Freakmighty, Freakofnurture, Fredrik, Fritzey97, Fryergreg, Furrykef, Gaius Cornelius, Giftlite, Gilliam, Glacialfox, Glen, Glenn, Globalsolidarity, Gogo Dodo, Golbez, Golgofrinchian, Gomada, GorillaWarfare, Graham87, Greenmango, GregorB, Greswik, Grstain, Gtg204y, Gtrmp, Guanaco, Gurch, Gwillhickers, H1nkles, Hackerrye, Halfway to never, Hamtechperson, Hanacy, Hanberke, Happysailor, Hapsiainen, Hdt83, Helpmet, HenryLi, Hevron1998, Hiddekel, Hiddenfromview, Hike395, Holoeconomics, HorsePunchKid, Hydrogen Iodide, Icairns, Immunize, Insanity Incarnate, Ioscius, Iridescent, Island Monkey, Ispy1981, J 1982, J appleseed2, J.delanoy, JDG, JForget, JFreeman, JHunterJ, JYolkowski, Ja 62, Jacob1207, Jadtnr1, Jake838, Jaksmata, JamieS93, Jantzen is cool, Jared Preston, JavierMC, Jiddisch, Jim.henderson, Jmrowland, Jn1611, Joefromrandb, Joelwest, John Cumbers, John Deming, John254, Jojhutton, Jon Harald Søby, Jonathunder, JorisvS, Jose77, Josh-Levin@ieee.org, Josh70597, Jpk, Jpo, Jschnur, Jusdafax, JustAddPeter, Jwrosenzweig, Jyril, Jóna Þórunn, Kazvorpal, Kborer, Keilana, Kelly Martin, Kenohki, KevWang, Kevyn, Kheider, Kier07, Kiiwiii, King of Hearts, Kingpin13, Kirill Lokshin, Kjoonlee, KleinKlio, Klilidiplomus, Kozawa, Kubigula, Kudret abi, Kurtan, Kuru, Kwamikagami, L337 kybldmstr, LALALA26, La Parka Your Car, Lamb99, Latics, Latitudinarian, Leevanjackson, Lesnail, Lexor, Lifefeed, Lightmouse, Lightyear69, Liken to a glove, LilliPuppy, Limonns, Ling.Nut, Logan, Lolihatepizza, Looxix, Loren.wilton, Loresayer, Lowe4091, Luckybeargod, Lugia2453, Luna Santin, Lupin, MATRIX, MER-C, MONGO, Macedonian, MacroDaemon, Maelnuneb, Magioladitis, Magister Mathematicae, Majorclanger, Mandarax, Mani1, Manmobile, MarcK, Marek69, Mark Foskey, Martarius, Martin451, Mas 18 dl, Masonbarge, Master Jay, Materialscientist, Mathewignash, Mattman00000, Matubo, Maury Markowitz, Mav, MavrickWeirdo, Mbc362, McSly, Mechanical digger, Mejor Los Indios, Menchi, Mendelssohn is Cool, Mephistophelian, Metalb, Metebelis, Mgmirkin, Michael Hardy, Micky kwoskie, Midnightcomm, Mike s, MikeCapone, MikeLynch, Mikenorton, Minesweeper, Minimac, MithrandirAgain, Mjuancarlo, Moomoomoo, Morwen, Moverton, Mrholybrain, Mtnerd, Muijz, Murgh, Murlough23, Mxn, Myanw, Mygerardromance, Myuphrid, NAHID, Nakon, Narcosis17, NawlinWiki, Neier, Nephelin, Nergaal, Netoholic, Newsaholic, Nick UA, Nihiltres, NiiSSA, Niobrara, Njaelkies Lea, Noclevername, Novangelis, Nubiatech, Ocaasi, Oleg Alexandrov, Olegwiki, Omicronpersei8, Orangemarlin, Oranjblud, OwenX, Oxymoron83, P. S. F. Freitas, PTJoshua, PTSE, Patanet7, Patmic2502, Patrick, Paul H., Paul Pot, Pauli133, Pedro, Perey, Pgan002, Phantomsteve, Philip Trueman, Pi@k, Pickom, Pinethicket, Piolinfax, Pit, Pizza Puzzle, PoeticVerse, Polkadotbaby0772, Polyamorph, Ponder, Porterjoh, Powered, Prashanthns, Pseudomonas, Python eggs, Q21op, Q43, Quadell, Quantpole, Quuxplusone, RFerreira, RJHall, RP88, RUL3R, RadicalOne, Rajabhattacharya, Ramesh Ramaiah, RandomCritic, Ravidreams, RazorICE, Razorflame, Rchandra, Reaper Eternal, Red Slash, RedWolf, Res2216firestar, Rettetast, RexNL, Reyk, Rich Farmbrough, Rivajunkie, Rjwilmsi, Rmhermen, RockMFR, Roentgenium111, Ron Ritzman, Rorschach, Rothorpe, Rubicon, Ruslik0, Rwflammang, Ryjaz, SJP, ST47, SU Linguist, Sagark86, Salmar, Sandhuharpreet, Sang'gre Habagat, Sapphic, Sardanaphalus, Saros136, Satori, SaveThePoint, Savemaxim, Saxsux, SchfiftyThree, Schneelocke, SchnitzelMannGreek, Sciurinæ, Sdornan, Senator Palpatine, Serendipodous, Serpentinite, Seth Ilys, Shadowjams, Shalom Yechiel, Shanes, Shirifan, Shotgun5559, Shushruth, SiliconCerebrate, SkerHawx, Skier Dude, Skizzik, SkyWalker, Slicedoranges, Slightsmile, Slon02, Snowolf, Soccer dude 07, Sodium, Sole Soul, Some jerk on the Internet, Something14, Soonhuat95, Southern Illinois SKYWARN, Sp82, Spacepotato, Spartaz, Spartenxiv, Spencer, Spitfire, Standunne, StarryGrandma, Stealingitstrongly, Stemonitis, Stephenchou0722, Steven Zhang, StitchRFC, Stormwriter, Sturmde, Stwalkerster, Subversive.sound, SummerPhD, SuperHamster, Superbeecat, Susan714, Sushi, Svetovid, Swatjester, Synchronism, Tailpig, Tassedethe, Taxman, Tbhotch, Telescopi, Template namespace initialisation script, That Guy, From That Show!, The Enlightened, The High Fin Sperm Whale, The Phoenix, The Singing Badger, The Thing That Should Not Be, TheGWO, TheKoG, Theda, Thestupidguy, ThreeBlindMice, Thue, Tide rolls, TigerShark, Tillwe, Tim123321Tim, Timir2, Timrollpickering, Timwi, Titoxd, Tjkiller555, Tom Peters, TomasBat, Tombomp, Tony Mach, Topgearstopviewer, TransUtopian, Trash boy, Tree Biting Conspiracy, Trekphiler, Tresiden, Tschild, Tyleradkins31, Ultra two, Uni223344, Untifler, Unyoyega, Uppland, Urhixidur, Vansh9988, Vary, Vaginator, Vgranucci, Vianello, Voxpuppet, Vrenator, Vsmith, Wackywace, Waltpohl, Waninge, Ward20, Wayne Hardman, Wayward, Werididiot, Werockthescienceclass, Westwood25, Wgungfu, Whitepaw, Why Not A Duck, Wiki alf, Wiki13, Wikiborg, Wikid77, Wikipelli, Wildyoda, Willhsmit, Willshark, WilyD, Wimt, Wipsenade, Wjfox2005, Wknight94, Wolfkeeper, WolfmanSF, Wwheaton, X relentless x, XJamRastafire, Xanucia, Xav71176, Xeno, Xerxes314, Xn44, Yamamoto Ichiro, Yoguys021298, Youknowandy, Yunshui, Z4ngetsu, Zidane tribal, Zsinj, Zundark, Zzzzz, ^demon, Île flottante, Саша Стефановић, Шугуан, 1441 anonymous edits

**Minor_planet** *Source*: http://en.wikipedia.org/w/index.php?title=Minor_planet *Contributors*: 3coma14, 7&6=thirteen, 786b6364, Aaron of Mpls, Aesopos, Alfons2, An Justified Wikipedian, Arsia Mons, Asikal1, Bmclaughlin9, Bryan Derksen, Chesnok, Ckatz, Colinmartin74, David Shepheard, Dicklyon, Eeekster, Fartherred, Fjörgynn, Glenn L, Gpermant, GraL, Interstellar Man, JHunterJ, Jolielegal, JorisvS, Kayau, Kheider, Kwamikagami, LeaveSleaves, MER-C, Max rspct, Merovingian, Michael Hardy, Miremare, Mrbenten, Murgh, Nergaal, Niobrara, Novangelis, Nubiatech, R'n'B, RJHall, Raymondwinn, Reyk, Rjwilmsi, Rmhermen, Roentgenium111, Romit3, Rorschach, Rothorpe, Ruslik0, Serendipodous, Sfahey, Shirt58, Sobreira, Soulkeeper, Spacepotato, Spidern, Terraflorin, The Tom, The way, the truth, and the light, Thingg, Thorncrag, Tlesher, Trekphiler, Ultra two, Urhixidur, Vaginator, Velpaedia Jenkuklordanus, West.andrew.g, Wikiprodder, Wiklol, William Avery, Willking1979, Woohookitty, WooperJeff, Writtenright, Wwannsda, 80 anonymous edits

**Dwarf_planet** *Source*: http://en.wikipedia.org/w/index.php?title=Dwarf_planet *Contributors*: (jarbarf), -Majestic-, 0030520dv, 1234567890Number, 21655, 68Kustom, 6SJ7, A, A-giau, AOB, AVM, Abtract, After Midnight, Ageekgal, AgentPeppermint, Ajraddatz, Alansohn, Alfredo.marelli, Ali@gwc.org.uk, Alsandro, Altenmann, Andonic, Andrew Gray, Anetode, Another Stickler, Anton Tudor, Anynobody, Aranae, Archola, Art LaPella, Arx Fortis, Astrazoic, Athaenara, Audiosmurf, AutomaticWriting, Avenue, BK520, Bakabaka, Banazir, Barbara Shack, Baseball

Watcher, Bayerischermann, Beland, Bender235, Benhocking, Bento00, Berek, Bfinn, Bflaminio, BiH, Bluap, BlueMoonlet, Bogatabeav, Bongwarrior, Brighterorange, Brisvegas, Bryan Derksen, Bsharitt, BudMann9, C'est moi, Calliopejen1, CanadianCaesar, CanadianLinuxUser, Canuckid, Catiline63, Catpowerzzz, Certh, Chairman S., Chochopk, Chr.K., Chris the speller, Circeus, Cjk91, Ckatz, Clicketyclack, Colinmartin74, Cometstyles, CovenantD, Csernica, Cuddlyopedia, Cyde, DVD R W, Damnfuct, DanPMK, Danogo, David136a, DeadEyeArrow, Deaþe gecweald, Delldot, Delldot on a public computer, DerHexer, Derek Balsam, Dfutato, Dgies, Dina, Discospinster, Doczilla, Dofarrell, Dojarca, DoktorDec, Doradus, Double sharp, Dr.Gonzo, DrKiernan, DragonflySixtyseven, Drc79, Dream out loud, Dubhe.sk, Dycedarg, DéRahier, E. Fokker, EamonnPKeane, Eccentricned, Ed Poor, El C, Eleusion, Eluchil404, Emurphy42, EoGuy, Epbr123, Epiphoney, Eurocommuter, Exodio, Extra999, FF2010, Fabartus, Face, FaerieInGrey, Fbs. 13, Fcn, Felix Portier, Figures&Puck, Filemon, Fjörgynn, FlieGerFaUstMe262, Frank12, Frankie1969, Frokor, Gareth Griffith-Jones, Gene Nygaard, Gene Ward Smith, George2001hi, Gfoley4, Ghalas, Giftlite, Givegains, Glacialfox, Glane23, Glenn L, Gnowor, Gordonrox24, Graldensblud, Grstain, Guliolopez, Hairy Dude, Harald88, Hayabusa future, Headbomb, Hede2000, Henry Flower, Heron, Hevron1998, Hibernian, Hillock65, Hmrox, Homo cosmosicus, Hon-3s-T, Horkana, Howcheng, IMSoP, IRP, Icairns, IcePuckScore, IdLoveOne, Ilmari Karonen, Incnis Mrsi, Ingolfson, Interstellar Man, Iridescent, Iridia, J.H.McDonnell, Jacobolus, Japanese Searobin, Jebba, Jeffrey Smith, Jerome Charles Potts, JoJaysius, JoeStanton, Johan1298, John Hyams, JohnCub, Jonwillig, Jopparn, Jor, JorisvS, Joseph Solis in Australia, Joshua BishopRoby, Jugni, Julian Mendez, JzG, Kanags, Katydidit, Kavkha, Keber, Kerotan, Kheider, Kirils, Kitch, Klingoncowboy4, Kmg90, Kmsiever, KnowledgeOfSelf, Koro Neil, Koyos, Kozuch, Krea, Kwamikagami, KyraVixen, LOL, LachlanA, Leandrod, Liberal Classic, Lightmouse, Liso, Lotje, Lueked, Lugia2453, MEBuckner, MK8, MONGO, MSGJ, MacTire02, Madonna Can, Majopius, Majorly, Malevious, Malo, Malzees, Man vyi, Mana Excalibur, Mandlhnyc, Maniesansdelire, Marcelo Pinto, Mareino, Markov, Martinwilke1980, MasterSci, Matdrodes, Materialscientist, MathEconMajor, MatthewMain, Matthewrbowker, Mckaysalisbury, Md84419, Mejor Los Indios, Menthaxpiperita, Metre01, Mfgreen, Mgmirkin, Michael Devore, Michael James Russo, Michaelas10, Michaelbusch, MicroProf, Mike Christie, Miniapolis, Miskatonic, MissAmbitious, Mkativerata, Montalto, Moochocoogle, MosheA, Moverton, Mpatel, Mrwuggs, Murgh, Murlough23, Mysdaao, Mzajac, Naddy, Nakon, Naohiro19, NawlinWiki, Nbbs, Nbound, Nergaal, NewEnglandYankee, Nickshanks, Nicolas Perrault III, Nicole1998, Nishkid64, Nixer, NuclearWarfare, Numbo3, Nuttycoconut, Nwbeeson, Nydas, O, One, Onorem, Opelio, Orangemarlin, Ouishoebean, Oxymoron83, P. S. F. Freitas, Parmaestro, Patricius Augustus, PatrikR, Patteroast, PedroPVZ, Peter Delmonte, Peter jackson, PeterSymonds, Petersburg, Pinethicket, Planemo, Plastikspork, Pmcray, PowerCS, Psiphiorg, Pulsfordp, QQQ, RP88, Rama, Ran, RandomCritic, RandomLittleHelper, Randomblue, Ravenswing, Readro, Recurring dreams, Redpants, RexNL, Reyk, Rgerhards, Rhillman, Riana, Rich Farmbrough, Richard B, RichardF, Rickyrab, Rigadoun, Rjwilmsi, Rmburkhead, Rmhermen, Roentgenium111, Rokkss, Rominandreu, Rossenglish, Rothorpe, RoyBoy, Rpyle731, Rst20xx, Rursus, Ruslik0, Ryulong, SP-KP, Salty!, Samuel Pepys, SandyGeorgia, Sardanaphalus, Saros136, Saturn star, Schgooda, Scimitar, Sciurinæ, Scjessey, Seddon, Seinfreak37, Sengkang, SenseiAC, Serendipodous, Sethhater123, Shadowmorph, Shamrockman455, Sharkface217, Shawnc, ShelfSkewed, Shii, ShutterBugTrekker, SidP, Siroxo, Sjakkalle, Skateboarderf2, Skizzik, SkyWalker, Slazenger, Sly Marbro 03, Soap, Solberger0127, Some jerk on the Internet, Something14, Sonjaaa, Sophia, Spangineer, Spitfire19, SplingyRanger, Steinbach, Steorra, Stephen123 2008, Stevan White, Steve2011, Stevebritgimp, Steven86, Stikonas, Stuntdawg, Summer noelle, Supersexyspacemonkey, Surfermoon, T.J.V., TTE, TXiKi, Tbayboy, Tbhotch, Teh Rote, Teh Xilian, TexasAndroid, Th1rt3en, The Enlightened, The Kid, The Singing Badger, The Tom, TheOtter, Thecheesykid, Tide rolls, Timberframe, Timc, TimothyHorrigan, Tomruen, TonyW, Treisijs, Trekphiler, Triona, Trypsin, Tsujigiri, Tycho Magnetic Anomaly-1, Tylerhammond3, Urhixidur, Vanka5, Vicki Rosenzweig, Vickyvicky2, Volland, Vyznev Xnebara, WDGraham, Wavelength, Wayne Hardman, WereSpielChequers, Wetman, Wighson, Wiki alf, Wild One, WinstonSmith, WolfmanSF, Woohookitty, Wrotesolid, Xerol, Xiahou, Yamamoto Ichiro, Yisraelasper, YorkBW, ZeekLTK, Zeimusu, Zfr, Zginder, Zntrip, 551 anonymous edits

**Near-Earth_object** *Source*: http://en.wikipedia.org/w/index.php?title=Near-Earth_object *Contributors*: A bit iffy, Africangenesis, Amberroom, Amrix, Anetode, Antonsusi, Anyeverybody, Arcayne, AstaG, Astral, Ataleh, Awickert, Baccyak4H, Bachrach44, Beaucouplusneutre, Beland, Benbest, Bender235, Bicanashy, Bporopat, Branlon, Breno, Bryan Derksen, Bubba73, Carcharoth, Cassini83, CesarB, Chesnok, Chrismichener1, CommonsDelinker, Conversion script, Coolgunz, Craigsjones, Cratylus3, Css, Curps, Czech is Cyrillized, Danim, Dark Tea, Darrenlatson, Darth Panda, Davidcannon, Dawnseeker2000, Denisarona, Dorftrottel, Dpenn89, Drttm, Dwlegg, Elonka, Epbr123, Eregli bob, Eric Kvaalen, Erisie, Euyyn, Everything counts, Fartherred, Fjörgynn, Fotaun, FrankTobia, Gaius Cornelius, Gene Nygaard, Gioto, Glenn, Gravitophoton, Green caterpillar, Greenrd, GregorB, Ground Zero, Grutness, Haim Berman, Hans Dunkelberg, Harryzilber, HopeSeekr of xMule, Hu, IanManka, Icairns, Ikluft, Imaginaryoctopus, Iohannes Animosus, Ixfd64, JDG, JIP, Jagons, Jeandré du Toit, Jeepo, Jngrossman, Joefromrandb, John Belushi, JorisvS, Joseph Solis in Australia, Kafziel, Kagemaru16, Kheider, Kinhull, Kwamikagami, LanceBarber, LarryGilbert, Larry_Sanger, Le sacre, Lionboy-Renae, Looxix, Lowe4091, Lunokhod, MATRIX, Manchurian candidate, Martijn Hoekstra, Matubo, Mav, Mdy66, Mets501, Michael Hardy, MikeCapone, Mikenorton, Mintleaf, Mitch Ames, Modest Genius, Mosesofmason, Mr Stephen, Mu301, Murgh, Muro de Aguas, Mzmadmike, N2e, NJR ZA, Neutrality, Ng.j, Nighend, NovaSilisko, Nr5278, Nurg, Oem2s3, Onsly, Oonissie, Oranjblud, Originalwana, PFSLAKES1, Patmic2502, Patrick, Paul H., PedroPVZ, Perry Middlemiss, Petersam, Phaldo, Phyrlight, Pohick2, Poodleboy, Primium mobile, PsuedoName, PuzzletChung, Q43, Qshio, RSStockdale, RedSpruce, Reyk, Rich Farmbrough, Rmhermen, Rockfang, Rod57, Roentgenium111, Rothorpe, RoyBoy, Rschweickart, Rst20xx, Ruslik0, Ryanjo, SAE1962, SMcCandlish, Sailbad, Saros136, Savidan, Science enthusiast343, Scotthatton, Sdsds, Serendipodous, Silence, Simulation18, Solipsist, SongScout, Sonicology, Spark010, Srschu273, Srwalden, SunSw0rd, TUF-KAT, Tbhotch, Terraflorin, Th1rt3en, The Anome, Thehotelambush, Trafford09, Trilobitealive, Twas Now, UltimateDarkloid, Under22Entreprenuer, Urhixidur, Uvaduck, Uvaphdman, Vadim Makarov, Wwheaton, Xenobiologista, Yankees76, Zafiroblue05, Zaqarbal, Zeimusu, 172 anonymous edits

**Centaur_(minor_planet)** *Source*: http://en.wikipedia.org/w/index.php?title=Centaur_%28minor_planet%29 *Contributors*: (, Acalamari, Alansohn, Argo Navis, Avinash.royyuru, Backin72, Beetstra, Bryan Derksen, CFLeon, Chesnok, Ckatz, CuriousEric, Curps, David Eppstein, Deuar, Dna-webmaster, Dread83, Duoduoduo, Egil, Eurocommuter, Filiosus's Saga, Fjörgynn, Fotaun, Gene Nygaard, Geoffrey.landis, Giftlite, Gurch, Gveret Tered, Hike395, Icairns, Ilmari Karonen, Iridia, Jmg38, John Hyams, JorisvS, Joseph Solis in Australia, Jscotti, Kbdank71, Kheider, Knowledge Seeker, Kwamikagami, Ling.Nut, Lowe4091, Maxim Razin, Mejor Los Indios, Mgmirkin, Michaelmas1957, Muma, Murgh, Njardarlogar, Novangelis, PamD, PedroPVZ, Piledhigheranddeeper, Pt, Rich Farmbrough, Rjwilmsi, Rmhermen, Roentgenium111, Ruslik0, Samdacruel, Sardanaphalus, Satori, Serendipodous, Serpentinite, Sethhater123, ShelfSkewed, Spartan, Tamfang, Template namespace initialisation script, The Singing Badger, The Tom, The shaggy one, Thebiggnome, Tom harrison, Tothebarricades.tk, Trafford09, Ufviper, Ulflarsen, Urhixidur, WolfmanSF, Δ, عبادة مجاهد ديرانية, 41 anonymous edits

**Trojan_(astronomy)** *Source*: http://en.wikipedia.org/w/index.php?title=Trojan_%28astronomy%29 *Contributors*: AHMartin, AdnanSa, All Is One, Amakuha, AndyKali, Baseballrocks538, BatteryIncluded, BilCat, Bryan Derksen, CSZero, Christopher Cooper, Creol, Curtis Clark, Dawnseeker2000, DerBorg, Drbogdan, Enchanter, Fjörgynn, Frankie1969, Frecklefoot, GTBacchus, Goustien, Grósznyó, Hairy Dude, Hans Dunkelberg, Hellbus, Jazzlvraz, JorisvS, Kevin Myers, Kjkolb, Kwamikagami, Lavallen, Leslie Mateus, Loadmaster, Mathias-S, Maurice Carbonaro, Medeis, Michael Hardy, Mimihitam, Mitch Ames, Njardarlogar, Ohms law, Piledhigheranddeeper, Qurq, RJHall, Rgwc, Rjwilmsi, Roentgenium111, Rothorpe, Ruslik0, Serasuna, Serendipodous, Skizzik, Spacepotato, TJRC, Tamfang, Tbhotch, Terraflorin, The Singing Badger, The Tom, TheStarmon, Theadept, Traxs7, Urhixidur, Westley Turner, Wkharrisjr, WolfmanSF, XavierGreen, Yworo, , 45 anonymous edits

**Solar_System** *Source*: http://en.wikipedia.org/w/index.php?title=Solar_System *Contributors*: (, (aeropagitica), -- April, -Majestic-, 129.128.164.xxx, 1297, 134340Goat, 141.157.52.xxx, 162.83.144.xxx, 216.237.32.xxx, 23skidoo, 24.93.53.xxx, 2nd Piston Honda, 711groove, A-giau, A. Parrot, AEMoreira042281, Aaron McDaid, Abbalkea2010, Abrech, Abuk78, Acalamari, Acather96, Acebrock, Acroterion, Adam850, Adashiel, Addshore, Adi, AdiJapan, Admiral Roo, Adraeus, Aelffin, Aeon1006, Ageekgal, AgentPeppermint, Ahadland1234, Ahoerstemeier, Ahsasin8, Aitias, Ajdz, Ajo Mama, Alanbly, Ale jrb, Alegoo92, AlexanderHaas, Alexdw, Algebraist, Alienkishore, Allstarecho, Alpertron, Altenmann, Amos Han, Amreatsf4620, Amyrlin, Anarchist42, Anaxial, Andonic, Andre Engels, Andrew Gray, AndrewHowse, Andrewlaneyboi93, Andy120290, AndyZ, Andyjsmith, Animum, Anitanorm, Anomalocaris, AnonEMouse, Antandrus, Anthony Arrigo, Anti-Quasar, AntiVan, Antiestablishment, Antonio Lopez, Anusface2000, Anyeverybody, Apers0n, Aquirata, Aragupta, Arbitrary username, Arce, Arctic.gnome, Ardric47, ArnoldReinhold, Art LaPella, Ashill, Ashleyy osaurus, Ashmoo, Asqwce, Aster2, Asteroidz R not planetz, Astrolog, Athenean, Ato, Atulsohan, Aubadaurada, Ausir, AutoFire, AutoGeek, AvicAWB, AxelBoldt, AySz88, AzaToth, Azcolvin429, BRG, BRUTE, Babbage, Badmonkey0001, Bajibble, Bambuush, BankingBum, Baseballdude12321, Batintherain, Bayerischermann, Bazzargh, Bbatsell, Bcorr, BeardedBob, Beast of traal, Beland, Belovedfreak, Ben Webber, Ben261994, BenAveling, Bender235, Benhocking, Bennymgradz, Berek, Bergsten, Bethy12, Beyazid, Bhadani, Bharathivdevi, Big Bird, Bigger digger, Bill37212, Billie Pipier, BirdKr, Bkdafkup, Bkell, Blakut, Blaxthos, Blaz5678, Blehfu, Blindman shady, Blobglob, BlytheG, Bob Hu, Bob rulz, Bobblewik, Bobo192, Boing! said Zebedee, Bomac, Bonadea, Bond81895, Bookworm2002, Boopolo, Bootstoots, BorgQueen, Borislav, Bowlhover, BozMo, Br77rino, Brian0918, Brianga, BrightStarSky, Brighterorange, Brion VIBBER, Brisvegas, Bruce1ee, Bruin69, Bryan Derksen, Bryan Seecrets, BryanG, BudMann9, Buddy777, Bugliarisi, Burrito, Burwellian, C0nanPayne, CWii, Cadiomals, Caid Raspa, Caltas, Can't sleep, clown will eat me, Canadian-Bacon, CanadianCaesar, Canderson7, Caravaca, CarbonUnit, Carlj7, Carnildo, Carultch, Catgut, Cderoose, Cenarium, Ceoil, Chairman S., Chantelle2004, Chaosdruid, CharlesC, Chem-awb, Chermundy, Chesnok, Chessphoon, Chipskip, Cholmes75, Chompzone, ChrisGriswold, Christopherlin, Chriswiki, Chupon, Ciaccona, Cimex, Cityside Seraph, Civil Engineer III, Ckatz, Clarityfiend, ClockworkTroll, Clq, Cmann100, Cmapm, Cmglee, Coastside, Cocoaguy, Cocu, Coelacan, Cogito ergo sumo, Cohenna, Coleflinn, ColinJF, Colinmartin74, Colonies Chris, Cometstyles, CommonsDelinker, Comp25, Complex (de), Computer user786, Computerjoe, Confession0791, Connelly, Constructive editor, Conversion script, Cooldude5298, Corpx, Costalotta, Cosumel, CovenantD, Coviepresb1647, Cquan, Crash Underride, Crazycomputers, Cremepuff222, Crevaner, Crevox, Cryptk, Cue the Strings, Curps, Cuttycuttiercuttiest, Cwill19, Cwolfsheep, Cyber boy, Cyclopia, Cyde, Cyktsui, Cymruman, CynicalMe, Cyril715, DARTH SIDIOUS 2, DDima, DMacks, DRTllbrg, DV8 2XL, Dabomb87, Daman numba23, Damërung, Daniel Olsen, Danielboerman, DannyZ, Danski14, Dar-Ape, Darkkid24, DarrynJ, DarthCat, DarthSwim, Dat789, Davewild, DavidLevinson, Davodd, Dazza anderson1, De Administrando Imperio, DeadEyeArrow, Dean Harper, Deaþe gecweald, Debresser, Deckiller, Decumanus, Deeptrivia, Dekaels, Dekimasu, Delaszk, Delfindakila, Delldot, Demmy100, Dendoi, DenisMoskowitz, Dennis Valeev, Denny, DerHexer, Derek Balsam, Derek Ross, Desmond71, Deuar, DeusMP, Deviathan, Dexeillic, Dia^, Diderot, Dina, Dirac1933, Discospinster, Dismas, Diverman, Dlimeb, Dlohcierekim, DmitryKo, DocWatson42, DocendoDiscimus, Dodo von den Bergen, Dogyo, Domthedude001, Don'tknowenough, Doradus, Dr. Submillimeter, DrKiernan, Dragons flight, Dragostanasie, Drat, Dreg743, Dreish, Drmies, Drumguy8800, Dry., Dummyacc, Duoduoduo, Dvandersluis, Dwaipayanc, Dwarfplanets.org.uk, Dwheeler, Dynamite12321, Dzordzm, E Pluribus Anthony, E Wing, EarthPerson, Ec5618, Ed de Jonge, Ed g2s, Ederiel, Edgar181, Editer in disguise cool, Edivorce, Eduardo Sellan III, Egyptianboy15223, Eiaschool, Eilthireach, El C, Elagatis, Elert, Elf, Elliskev, Emarsee, Emilio floris, Emokid0987654321234567890, Ems57fcva, Emufarmers, Enceladusgeysers, Enirac Sum, Enok, Enviroboy, Eob, Epbr123, Equal, Er Komandante, Erebus555, Ergzay, Eric Forste, Eric Wester, Erikhansson1, Erimus, Eroica, Esanchez7587, Etxrge, Evanspahr5, Evercat, Everyking, Evgeni Sergeev, Evil saltine, Exert, Explicit, Extra999, Extransit, FF2010, Facts707, Fanghong, Farosdaughter, Farry, FastLizard4, Fatladi, FayssalF, Fedayee, Felix Portier, Ferdiaob, Feydey, Feyrauth, Figures&Puck, Fjörgynn, Flata, Fluorhydric, Flyguy649, Fnfd, Foobar, Fotaun, Franamax, Frank Shearar, Freakofnurture, Fredrik, Fredvanner, Freewayguy, Freiza667, FreplySpang, Frip1000, Fullerene, Futuristcorporation, GDonato, Gadfium, Gaijin Ninja, Gaius Cornelius, Galoubet, Gameseeker, Gary King, Garycompugeek, Gateman1997, Gdo01, Geckodude92, Gene Mocoletlet, Gene Nygaard, GenestealerUK, Geni, Geob42, Geoff.green, Geologyguy, Georgia guy, Gerbrant, Ghewgill, Gian-2, Giftlite, Gifðas, Giggy, Gilliam, Gilsinan, Gimboid13, Glass Sword, Glen, Glenn, Glenn L, Gloern, Gman124, Gnixon, Gnomon Kelemen, GoShow, Gogo Dodo, GoingBatty, Golbez,

GoldenTorc, Gone999, Goodnightmush, Gorillaman0321, Gota 93, GraemeL, Grandrek, Greenmage801, Greentsunami, Gregconquest, Grika, Ground Zero, Grstain, Grunt, Gscshoyru, Gsklee, Gt, Gurch, Gurchzilla, Guyzero, Gwaldo, Gwillhickers, H, H2g2bob, Hadal, Hakkahakkabazoom, HalfShadow, Halsteadk, HammerHeadHuman, Harperska, HarryAlffa, Harryboyles, Harsh 2580, Hayter, Haza-w, Hdt83, Headbomb, Heals33, Hello i am a guy, Henhellen, Hermitage17, Heron, Hersfold, Hertz1888, Hevron1998, Hi332211, HiDrNick, HiLo48, Hibernian, Hike395, Hmartin, Hojimachong, HorsePunchKid, HumphreyW, Hurricane Devon, Husond, Hut 8.5, Hyad, Hypercephalic, I know about this, I'm supah awesome, ILHI, Iamunknown, Ianblair23, Iantresman, Ibagli, Ibroadbent, Icairns, IcePuckScore, Icestorm815, IdLoveOne, Idaltu, Ideogram, Iliev, Ima van Daal, ImhotepBallZ, Insanity Incarnate, InternetMeme, Iridescent, Irishguy, Iron knife, IronGargoyle, Isoxyl, Ixfd64, J P, JBG, JBlover23, JD554, JFreeman, JHunterJ, JQF, JStarStar, JWSchmidt, Jagged 85, James D. McBride, Jan.Kamenicek, Janviermichelle, Jao, Japak, Jaranda, Jason One, Jason Palpatine, JeLuF, Jeandré du Toit, Jeffmedkeff, Jeremy Visser, Jeronimo, Jerry, Jevans14, Jfiling, Jfpierce, Jfullerw, Jhenderson777, Jim Fitzgerald, Jim bob jr, Jimp, Jinma, Jkelly, JoanneB, JodyB, Joeborder, Joelwest, Joffan, John Quincy Adding Machine, JohnCub, JohnDBuell, Johnismyboo, Johnuniq, Jojit fb, JonWee, Jonathan Drain, Jorfer, JorisvS, Joseph Solis in Australia, Josh Parris, Josh Tumath, Joshua368, Jpallan, Jpk, Jredmond, Jrethorst, Jrockley, Julesd, Jumbuck, Jusjih, JustAGal, Jyril, K, K-UNIT, KAMiKAZOW, Kaldari, Kanags, Karaku, Karozoa, KathrynLybarger, Katydidit, Kaustubh, Kayau, Kchishol1970, Kcordina, Keilana, Kelvinc, Keraunos, Kevin29303, Keyboard mouse, Khalid Mahmood, Kheider, Kiensvay, Kim Bruning, King Lopez, Kingturtle, Kirbytime, Kitch, Kjkolb, Kjoonlee, Kmsiever, Knowledge Seeker, KnowledgeOfSelf, KojieroSaske, Kortaggio, Kozuch, KrakatoaKatie, Krash, Krich, Krótki, Kubigula, Kukini, Kumpees gay, Kungfuadam, Kunguni, Kuru, Kush2, KuwarOnline, Kwamikagami, Kwekubo, Kyokpae, KyraVixen, L Kensington, LAX, LOL, La goutte de pluie, LachlanA, Lacrimosus, Lanthanum-138, Laserfire, Latitude0116, Laudaka, Leafyplant, Lectonar, Lee, Leevanjackson, Leevclarke, Legolost, Leigui, Leki, Leon7, Lethe, Lexicon, LifeStar, Liftarn, Lightmouse, Lights, Limbo socrates, Linnell, Liop666, Liyster, Lo2u, Lomn, LonelyMarble, Lookang, Looxix, Lord Patrick, Lordmontu, Love fass, Lradrama, LuckyBob76, Luckyluke, LuigiManiac, Lukaigeneration, LukeSurl, Lulujannings, Lumos3, Luna Santin, Lunokhod, Lupin, MER-C, MKar, MONGO, MacTire, Madgeek09, Madhav krishnan, Madonna Can, Magister Mathematicae, Majorly, Makron1n, Malo, ManCreamy, Manipe, Manisero399, Manjithkaini, MaraNeo127, Marcus Qwertyus, Mark Foskey, Mark Musante, MarkSutton, Markhurd, MarsRover, Marshallsumter, Martin benettis@hotmail.com, Martin.Budden, Mastalroh, Master Jay, MasterSci, Masterhomer, Materialscientist, Matt Crypto, Matt. P, Matthead, Matty456789, Maurreen, Mav, Maxim, Mayumashu, Mboverload, Mbz1, McLarenJAB, Mdd, Mdf, Megabyte73, Meidosemme, Melaen, Memaster3, Meneth, Meno25, MeowMedia, Merovingian, Mets501, MetsFan76, Mfrontz, Michael Hinrichs, Michael Patrick, Michael Slone, Michael riber jorgensen, Michaelas10, Michaelbarreto, Michaelbusch, Michalis Famelis, Microtony, Midgrid, Midnightcomm, Mightymights, Mike Christie, Mike Doughney, Mike s, Mikebrand, Mild Bill Hiccup, Mimihitam, Minesweeper, Minna Sora no Shita, MisfitToys, Misheru, MisterSheik, Mjb, Mlm42, Moanzhu, Mobiusorg, Mode11bob9, Monedula, Monotonehell, Montana's Defender, Monz, MoonshineOHNO, Mooseman927, Morelanj, Mormegil, Morshem, Moverton, Mpatel, Mr Tan, Mr. Garrison, Mr. Lefty, Mr. Vernon, Mred64, Mrexxx, Mschel, Msikma, Mstroeck, MuDavid, Mufka, Multifinder17, Munchynik, Murgh, Murlough23, Mwtoews, Mxn, Myanw, Myguy123, Myles Trundle, Myrrhlin, Mzajac, N-Man, Nabla, Nacnud298, Nadav12, Nakon, Nations United, NatureA16, NawlinWiki, Nbound, Nedim Ardoğa, Nelson50, Nemrac2, Neonumbers, Nephelin, NerdyScienceDude, Nergaal, Nesnad, Neutrality, Neverquick, Newone, Nfitz, NickBush24, NickW557, Nickshanks, Nigholith, Night Gyr, Nihiltres, Nikai, Nimur, Nivix, Nixer, No-Bullet, NoIdeaNick, NobodysPerfect312, Noctibus, Noetica, Noosfractal, NorCalHistory, Norn Guy, Northumbrian, Notinasnaid, Nova77, Novangelis, Nurg, Nuttycoconut, Nxavar, Oboebasson, Ohms law, Olathe, Oliphaunt, Oliver Pereira, Oliver202, Omicronpersei8, Once in a Blue Moon, Onco p53, One, Onorem, Opelio, Orangutan, OrbitOne, Oren0, Osmaston, Ottershrew, Ouishoebean, Oxymoron83, P. S. F. Freitas, PFHLai, PJS102, PJtP, PMP126, Paine Ellsworth, Paliosun, Panoptical, Papajfd, Parachnidox, Parsa, Patrick, Patrickwooldridge, Patstuart, Paul August, Paul-L, Paulcmnt, Paulkondratuklovesny, Pb30, Pbrower2a, Pcj, Peash, Pedro, Penubag, Pergamino, Perrysweitzer, Peruvianllama, Peter, Peter Shearan, Peter Winnberg, Peterfwhyte, Pgecaj, Pgk, Pgr94, Ph89, Pharaoh of the Wizards, Phe, Phil 1970, Philip Trueman, Phillip J, Philolexica, Phoenix2, Phoenix79, PhySusie, Pi@k, Pictureuploader, Pilatus, Pilotguy, Pinethicket, Pious7, Piper2000ca, Pizza Puzzle, Pjrm, PlanetStar, Planetary, Plau, PoccilScript, Poiu555, Polly, Polonium, Poor Yorick, Poplard, Postmortemjapan, Potatoedrink, PresN, Primetime, Prodego, Prometheus7Unbound, Proxima Centauri, Psyche825, Puntori, PuzzletChung, Pyrokitsune777, Python eggs, Q2op, Q3op, Qmwne235, QuantumEleven, Quintote, Qurq, Qwasty, Qwyrxian, Qxz, R, RJHall, RP88, Rachel Pearce, RadicalPi, Raditzu, Raichu Trainer, Raistlin11325, Raj Fra, Rake, Rama's Arrow, RandomCritic, Randomblue, Rashaani, Rasmus Faber, Ratsbew, Raven4x4x, Raymondwinn, RazorICE, Rc3784, Real editor, Redgolpe, Redsully, Reedbraden, Renesis, Rettetast, RexNL, Rhuaridh, Riana, Richard D. LeCour, RichardF, Rick Block, Rickyrab, Ringmaster56, RingtailedFox, Rizzoj, Rjwilmsi, Rmky87, Rnt20, RobHolbert, Robbie098, Robby, RobertG, Robomaeyhem, RockMFR, Roentgenium111, Roerlepel, RoflSloth, Ronhjones, Roo60, Roofus, Rorschach, RoryReloaded, Roscoe x, Rothorpe, RoyBoy, Rreagan007, Rror, Rubicon, Run!, Rune.welsh, Rursus, Ruslik0, Ruud Koot, Rwboa22, Ryan Paddy, Ryulong, SJP, SMC89, SOCCERchic2134, SP-KP, ST47, Sabrebd, Sacre, Salamander44, Salamander724, Salamurai, Saluyot, Salvio giuliano, Sam Korn, Sam12321, Samjohnson, Samojari, Samoojas, Sampo Torgo, Samy Merchi, Sanchom, Sanderling, Sango123, Saramae101, Sardanaphalus, Saros136, Savant13, Sbandrews, Scarvie, Schneelocke, SchuminWeb, Schzmo, Scienz Guy, Scipius, Sciurinæ, Scjessey, Scooter20, Scott Ritchie, Scottosborne, Scwlong, Sd31415, Sean D Martin, Sean higgy2007, Selket, Sengkang, Sennaista, Serendipitous, Serendipodous, Sfahey, Sgt Pinback, Shadow007, Shadowjams, Shanes, Shanew2, Shawn81, SheffieldSteel, Shehee, Sherool, Sherz, Shinjiman, Shoeofdeath, Sidasta, Sietse Snel, Sikon, Silivrenion, Simple implication, Siniset, Sinn, Sir Nicholas de Mimsy-Porpington, Siroxo, Sj, Sjakkalle, SkerHawx, Skizzik, SkyMachine, Slakr, Sleddog116, Sleeping123, Slysplace, Smartech, Smartse, Smarty Pants786, Smige30, Snigbrook, Snowolf, Solipsist, Someguy1221, Something14, Soumya11, Soumyasch, Spaceman8815, Spacepotato, Spangineer, Sparky the Seventh Chaos, SparrowsWing, Spazturtle, Speed Air Man, Spellcast, Spiral Wave, SpookyMulder, SpuriousQ, Squigish, Sry85, Stan Shebs, Starexplorer, Statalyzer, Stattouk, Stefan, Stellmach, StephenDawson, Stephenb, SteveMcCluskey, Steven J. Anderson, Steveo2, StoptheDatabaseState, Stwalkerster, Summabrian, Sun Creator, Sundar, Supabanakid, SuperExpress, Superm401, Supermat, Supernova0, Surachit, Svdragunov, Sweetness46, Swpb, Symo85, Syndicate, Szajci, Tails Soda, Tamfang, Tangotango, Tanzania, Tariqabjotu, Tarotcards, Tarquin, Tarquin Binary, Tassedethe, Tastycheeze, Tbayboy, Tdadamemd, Tdvance, TeaDrinker, Teh Xilian, Template namespace initialisation script, Terminal157, TexasAndroid, Th1rt3en, The Enlightened, The Gaon, The High Fin Sperm Whale, The Infidel, The Merciful, The Phoenix, The Placebo Effect, The Rambling Man, The Singing Badger, The Thing That Should Not Be, The Tom, The shaggy one, TheCodeman4, TheTruth2, Theda, Thedjatclubrock, Theopolisme, Thierry Caro, Thomasedavis, Thorpe, Thunderbrand, Thw1309, Tianxiaozhang, Tideflat, Tigah, TimShell, Time3000, Timir2, Timwi, Tinctorius, Tipou, Titoxd, Tkbwik, Tktktk, Tlogmer, Tom, Tommi Ronkainen, Tompw, Tomruen, Tonic123, Tony1, Topbanana, Tothebarricades.tk, Tpbradbury, TransistorsGoneWild, Traroth, Travelbird, Treybien, Tricky Victoria, Trottaman471, Trunks6, Trusilver, Tslocum, Ttsenis, Tulkolahten, Turnteeth, Turok 11, Turok 12, Turok 16, Turok 5, TutterMouse, Tuvas, TxMCJ, Tycho Magnetic Anomaly-1, Tyler148, Tyr Anasazi, U.S.A.U.S.A.U.S.A., UberScienceNerd, Umlautbob, UncleBubba, Uniqueuponhim, Urhixidur, Us441, Ushionna, Vanished user 5zariu3jisj0j4irj, Vary, Vasi, Vedranf, Vicki Rosenzweig, Viridian, VoidLurker, Voldemort, Voortle, Vowlesy, Voxii, Vrenator, Vsmith, Vy0123, Vyznev Xnebara, Watch37264, Wavelength, Wayward, Wdfarmer, Weatherman1126, Wenli, Wenny888, Werdan7, Werothegreat, Weyes, Whatismetric, Whcernan, Who, Whoop whoop pull up, WikHead, Wiki alf, Wikianon, Wikicheng, Wikiwatchers, Will.i.am, WillMak050389, Wilsjrwinners, Wimt, WingZero, Wink wink, Wise1990, Wisq, Wknight94, Wmahan, Wolfkeeper, WolfmanSF, Worldtraveller, XJamRastafire, Xammer, Xanzzibar, Xenonice, Xession, Xiahou, Yamamoto Ichiro, Yelyos, Yintan, Yodaroxmysox, Yupik, Yuya takeuchi, Zafiroblue05, Zanaq, Zaphodthewise, Zbayz, Zondor, Zowie, Zsinj, Zvika, Zzuuzz, Zzyzx11, Ævar Arnfjörð Bjarmason, Τις, Товарищ, ئارس نورى، باس‌مه , 2399 anonymous edits

**Kuiper_belt** *Source*: http://en.wikipedia.org/w/index.php?title=Kuiper_belt *Contributors*: @pple, A2-computist, Aajacksoniv, Aaron of Mpls, AbbaIkea2010, Acer, Acom, AdSR, Aditya, Afroman123456789, Alfio, Altenmann, Amaurea, Andre Engels, Andres, Another Believer, Arctic.gnome, ArgGeo, Art LaPella, AshLin, Astor14, AstroHurricane001, AstroMark, AstroNomer, Ataleh, Audriusa, Av0id3r, Avsa, Barneca, Bender235, Berek, BertSen, Bfigura's puppy, Bigger digger, BlueMoonlet, Bobo192, Bolo1729, Bongwarrior, Borgx, Boston, Brandmeister, Brian0918, Brighterorange, Brim, Brion VIBBER, Bryan Derksen, Bubba73, C777, CWenger, Caerwine, Cam, Carbuncle, Card, Chesnok, Christopher Cooper, Christopher Parham, CityOfSilver, Closedmouth, Co149, Cobblet, Cometstyles, Conorobradaigh, Conversion script, Cookiehead, Craig Pemberton, Ctachme, CuriousEric, Curps, Cwolfsheep, Cyberia23, Cyp, Cyrius, DNewhall, Dabomb87, Dan Austin, Danny, Deathphoenix, Deflective, Denelson83, Desibites, Deuar, Dhum Dhum, Dna-webmaster, Downhighest, Drvancampen, Ec5618, Ed Cormany, Egil, Eippiepie33, Eleassar777, ElectricValkyrie, Ellywa, Elockard, Eob, Epbr123, Eric Wester, Ericg, Escape Orbit, Eurocommuter, Evercat, Evil Monkey, Exir Kamalabadi, FF2010, Fartherred, Fenneth, Firsfron, Fivemack, Fjörgynn, Fluorhydric, Foosher, Fredrik, GabrielEvans, Garglebutt, Geeoharee, George Ponderevo, Giftlite, Gilliam, Glenn, Gnickett1, GoingBatty, Graham87, Grtamlinb, Guessing Game, Hairy Dude, Harold f, Harp, HarryAlffa, Harryboyles, Headbomb, Heron, Hike395, Howcheng, HumphreyW, Iamunknown, Ignoranteconomist, Ilmari Karonen, Imgayftwyes, Inge-Lyubov, Iridescent, Iridia, Ishel99, J P, JDT1991, JMK, JamesFox, Jan.Kamenicek, Jann, Jaredisred, Jarry1250, JasonAQuest, Jasongetsdown, Jay Litman, Jbaugher13, Jeandré du Toit, Jeffmedkeff, Jeronimo, Jk2693, Joedeshon, Joelwest, John, John Belushi, John Hyams, John85, Jolielegal, Jon.prasad, Jonshill, Jor, JorisvS, Josh Grosse, Josiah Rowe, Jyril, KBi, Kanags, Keflavich, Kevyn, Khalid Mahmood, Kheider, Kitsunegami, Kiviuq, Kozuch, Kwamikagami, Latulla, Lestatdelc, Lightblade, Ling.Nut, Logan, LonelyMarble, LonelyPker, Looxix, Lowe4091, Luk, Lusanaherandraton, Lwalt, MK, MK8, MONGO, Madkayaker, Malhonen, Martin56, Maurice Carbonaro, Mav, Mbell, Mdotley, Mendaliv, Mentifisto, Michael Devore, Michael Hardy, Mike s, Mikegrant, Minesweeper, Modest Genius, Monz, Mooquackwooftweetmeow, Moritheil, Mrwuggs, MuZemike, Murgh, Murlough23, Myrrhlin, Mzajac, Nascarfan1,000,000,000, NeilTarrant, Nemo1986, Nemu, Nephelin, Ngebendi, Nickshanks, Nixer, Njardarlogar, Ocee, Odros, Ohconfucius, Olivier, Omnedon, OrgasGirl, Originalwana, Ottawa4ever, Outriggr, Ozone77, P. S. F. Freitas, P.T. Aufrette, Pablo-flores, Padalkar.kshitij, Palica, Pauli133, Peashy, Pharos, Philip Trueman, Philrosenberg, Piast93, Piledhigheranddeeper, Pizza Puzzle, Pmokeefe, Ponder, Poop face3432, Poppy, Princejack12, RA0808, RMHED, RP88, RadicalOne, Rasimpson, Rathanor, Rebrane, RedBLACKandBURN, RexNL, Rhodekyll, Rhort, Rich Farmbrough, Richard B, Richhoncho, Rickyrab, Rjwilmsi, Robert Merkel, RobertG, Roentgenium111, Ronhjones, RoosMargot, RossPatterson, Rothorpe, Roudy66, RoyBoy, Runningonbrains, Rursus, Ruslik0, Ruy Pugliesi, SHCarter, Sage of Ice, Sardanaphalus, Sardur, Saros136, Schneelocke, SchuminWeb, Scjessey, Scog, ScottyBoy900Q, Scruffywiki, Scwlong, Semperf, Serendipodous, Sffcorgi, Shadowjams, Shadowmorph, Shawn81, Shirulashem, Shsilver, Sich Bojan, Sideways713, Sietse Snel, Sintaku, Sir Nicholas de Mimsy-Porpington, Smartech, Some jerk on the Internet, Something14, Spellmaster, StephenBuxton, Stepheng3, Stevenj, Stirling Newberry, Sturmde, T-borg, TTE, Tamfang, Tarotcards, Tedzsee, Tegles, Template namespace initialisation script, TeunSpaans, The Other Saluton, The Thing That Should Not Be, The Tom, The Yeti, TheMadBaron, TheMightyOrb, Thegreatdr, Tide rolls, TimeCruiserMike, Timwi, Tom, TomXP411, Tomchiukc, Treisijs, Triplepickle815, Triwbe, Tronno, Tuvas, Tycho Magnetic Anomaly-1, Ufviper, Uncle G, Upquark, Urhixidur, Vahagn Petrosyan, Vanis314, Vegeta624, Vicki Rosenzweig, Vimalkalyan, Vuong Ngan Ha, Wayne Hardman, Wiki alf, Wikipelli, WilyD, WolfmanSF, Writtenonsand, Wysprgr2005, XJamRastafire, XLerate, Xiahou, Ylee, Yogirox234, Yosri, Zad68, Zero sharp, Zhatt, Саша Стефановић, باس‌مه ₒx, 444 anonymous edits

# Image Sources, Licenses and Contributors

**Image:InnerSolarSystem-en.png** *Source*: http://en.wikipedia.org/w/index.php?title=File:InnerSolarSystem-en.png *License*: unknown *Contributors*: Original uploader was Mdf at en.wikipedia

**Image:Giuseppe Piazzi.jpg** *Source*: http://en.wikipedia.org/w/index.php?title=File:Giuseppe_Piazzi.jpg *License*: unknown *Contributors*: Howcheng, Lalupa, Lotse, Papa November, Rentier, Rootology, Ruslik0

**Image:Main belt i vs a.png** *Source*: http://en.wikipedia.org/w/index.php?title=File:Main_belt_i_vs_a.png *License*: unknown *Contributors*: Piotr Deuar

**Image:951 Gaspra.jpg** *Source*: http://en.wikipedia.org/w/index.php?title=File:951_Gaspra.jpg *License*: unknown *Contributors*: Alaniaris, BokicaK, Papa November, Ruslik0, The Evil IP address, Tryphon, Yonidebest, 5 anonymous edits

**Image:AllendeMeteorite.jpg** *Source*: http://en.wikipedia.org/w/index.php?title=File:AllendeMeteorite.jpg *License*: unknown *Contributors*: User:Vesta

**Image:Main belt e vs a.png** *Source*: http://en.wikipedia.org/w/index.php?title=File:Main_belt_e_vs_a.png *License*: unknown *Contributors*: Original uploader was Deuar at en.wikipedia

**Image:Kirkwood Gaps.svg** *Source*: http://en.wikipedia.org/w/index.php?title=File:Kirkwood_Gaps.svg *License*: unknown *Contributors*: based on plot by Alan Chamberlain, JPL/Caltech

**Image:zodiacal.jpg** *Source*: http://en.wikipedia.org/w/index.php?title=File:Zodiacal.jpg *License*: unknown *Contributors*: Original uploader was Serendipodous at en.wikipedia

**Image:Asteroid proper elements i vs e.png** *Source*: http://en.wikipedia.org/w/index.php?title=File:Asteroid_proper_elements_i_vs_e.png *License*: unknown *Contributors*: Bryan Derksen, Deuar, Papa November, Ruslik0, WinstonSmith, Yann, 1 anonymous edits

**Image:Dawn Flight Configuration 2.jpg** *Source*: http://en.wikipedia.org/w/index.php?title=File:Dawn_Flight_Configuration_2.jpg *License*: unknown *Contributors*: GDK

**Image:Asteroidsscale.jpg** *Source*: http://en.wikipedia.org/w/index.php?title=File:Asteroidsscale.jpg *License*: unknown *Contributors*: NASA/JPL-Caltech/JAXA/ESA

**Image:4 Vesta 1 Ceres Moon at 20 km per px.png** *Source*: http://en.wikipedia.org/w/index.php?title=File:4_Vesta_1_Ceres_Moon_at_20_km_per_px.png *License*: unknown *Contributors*: NASA, composition by Urhixidur

**Image:Ceres symbol.svg** *Source*: http://en.wikipedia.org/w/index.php?title=File:Ceres_symbol.svg *License*: unknown *Contributors*: Aquantrum, Crei0, Galphi, Lexicon, Ramiy, Ruslik0, Sarang, Sarilho1, Urhixidur, 4 anonymous edits

**Image:Ceres2.svg** *Source*: http://en.wikipedia.org/w/index.php?title=File:Ceres2.svg *License*: unknown *Contributors*: Varun Kumar

**Image:Ceres3.svg** *Source*: http://en.wikipedia.org/w/index.php?title=File:Ceres3.svg *License*: unknown *Contributors*: Varun Kumar

**Image:2Pallas symbol.svg** *Source*: http://en.wikipedia.org/w/index.php?title=File:2Pallas_symbol.svg *License*: unknown *Contributors*: User:KleinKlio

**Image:2 Pallas.svg** *Source*: http://en.wikipedia.org/w/index.php?title=File:2_Pallas.svg *License*: unknown *Contributors*: Vectorized by

**Image:Juno symbol.svg** *Source*: http://en.wikipedia.org/w/index.php?title=File:Juno_symbol.svg *License*: unknown *Contributors*: Lexicon, Sarang, Urhixidur

**Image:3 Juno (1).svg** *Source*: http://en.wikipedia.org/w/index.php?title=File:3_Juno_(1).svg *License*: unknown *Contributors*: User:Sim0nx

**Image:Symbol 3.jpg** *Source*: http://en.wikipedia.org/w/index.php?title=File:Symbol_3.jpg *License*: unknown *Contributors*: Bkell, Garion96, Pickom, RP88

**Image:Vesta symbol.svg** *Source*: http://en.wikipedia.org/w/index.php?title=File:Vesta_symbol.svg *License*: unknown *Contributors*: G.dallorto, Lexicon, Rmhermen, Sarang, Urhixidur

**Image:4 Vesta (0).svg** *Source*: http://en.wikipedia.org/w/index.php?title=File:4_Vesta_(0).svg *License*: unknown *Contributors*: Roland Geider

**Image:Simbolo di Vesta.svg** *Source*: http://en.wikipedia.org/w/index.php?title=File:Simbolo_di_Vesta.svg *License*: unknown *Contributors*: Original uploader was Kwamikagami at en.wikipedia

**Image:4 Vesta Unsimplified Symbol.svg** *Source*: http://en.wikipedia.org/w/index.php?title=File:4_Vesta_Unsimplified_Symbol.svg *License*: unknown *Contributors*: Original uploader was Tooto at en.wikipedia

**File:5 Astraea symbol alternate.svg** *Source*: http://en.wikipedia.org/w/index.php?title=File:5_Astraea_symbol_alternate.svg *License*: unknown *Contributors*: Wilson, John

**Image:5 Astraea Symbol.svg** *Source*: http://en.wikipedia.org/w/index.php?title=File:5_Astraea_Symbol.svg *License*: unknown *Contributors*: User:Rmhermen

**Image:6 Hebe Astronomical Symbol.svg** *Source*: http://en.wikipedia.org/w/index.php?title=File:6_Hebe_Astronomical_Symbol.svg *License*: unknown *Contributors*: User:Rmhermen

**Image:7 Iris Astronomical Symbol.svg** *Source*: http://en.wikipedia.org/w/index.php?title=File:7_Iris_Astronomical_Symbol.svg *License*: unknown *Contributors*: Editor at Large, Ruslik0, Urhixidur, Wikiborg

**Image:8 Flora Astronomical Symbol.svg** *Source*: http://en.wikipedia.org/w/index.php?title=File:8_Flora_Astronomical_Symbol.svg *License*: unknown *Contributors*: Editor at Large, Urhixidur

**Image:9 Metis symbol.svg** *Source*: http://en.wikipedia.org/w/index.php?title=File:9_Metis_symbol.svg *License*: unknown *Contributors*: User:Bryan Derksen

**File:10 Hygeia symbol alternate.svg** *Source*: http://en.wikipedia.org/w/index.php?title=File:10_Hygeia_symbol_alternate.svg *License*: unknown *Contributors*: Hind, J. R.

**Image:10 Hygiea Astronomical Symbol.svg** *Source*: http://en.wikipedia.org/w/index.php?title=File:10_Hygiea_Astronomical_Symbol.svg *License*: unknown *Contributors*: User:Editor at Large

**File:11 Parthenope symbol alternate.svg** *Source*: http://en.wikipedia.org/w/index.php?title=File:11_Parthenope_symbol_alternate.svg *License*: unknown *Contributors*: Wilson, John

**Image:11 Parthenope symbol.svg** *Source*: http://en.wikipedia.org/w/index.php?title=File:11_Parthenope_symbol.svg *License*: unknown *Contributors*: User:Editor at Large

**Image:12 Victoria symbol.svg** *Source*: http://en.wikipedia.org/w/index.php?title=File:12_Victoria_symbol.svg *License*: unknown *Contributors*: User:Editor at Large

**File:13 Egeria symbol.svg** *Source*: http://en.wikipedia.org/w/index.php?title=File:13_Egeria_symbol.svg *License*: unknown *Contributors*: Hind, John Russell

**File:14 Irene symbol.svg** *Source*: http://en.wikipedia.org/w/index.php?title=File:14_Irene_symbol.svg *License*: unknown *Contributors*: John Wilson

**Image:15 Eunomia symbol.svg** *Source*: http://en.wikipedia.org/w/index.php?title=File:15_Eunomia_symbol.svg *License*: unknown *Contributors*: User:Bryan Derksen

**File:16 Psyche symbol.svg** *Source*: http://en.wikipedia.org/w/index.php?title=File:16_Psyche_symbol.svg *License*: unknown *Contributors*: J. R. Hind

**Image:17 Thetis symbol.png** *Source*: http://en.wikipedia.org/w/index.php?title=File:17_Thetis_symbol.png *License*: unknown *Contributors*: Symbol created by R. Luther

**Image:18 Melpomene symbol.svg** *Source*: http://en.wikipedia.org/w/index.php?title=File:18_Melpomene_symbol.svg *License*: unknown *Contributors*: Hind, John Russell

**File:19 Fortuna symbol.svg** *Source*: http://en.wikipedia.org/w/index.php?title=File:19_Fortuna_symbol.svg *License*: unknown *Contributors*: Hind, John Russell (May 12, 1823 – December 23, 1895)

**Image:26 Proserpina symbol.svg** *Source*: http://en.wikipedia.org/w/index.php?title=File:26_Proserpina_symbol.svg *License*: unknown *Contributors*: Symbol created by R. Luther

**Image:28 Bellona symbol.svg** *Source*: http://en.wikipedia.org/w/index.php?title=File:28_Bellona_symbol.svg *License*: unknown *Contributors*: User:Editor at Large

**Image:29 Amphitrite symbol.svg** *Source*: http://en.wikipedia.org/w/index.php?title=File:29_Amphitrite_symbol.svg *License*: unknown *Contributors*: Chauncey A. Goodrich and Noah Porter, editors

**Image:35 Leukothea symbol.png** *Source*: http://en.wikipedia.org/w/index.php?title=File:35_Leukothea_symbol.png *License*: unknown *Contributors*: G. Rumker and C.A.F. Peters

**Image:37 Fides symbol.svg** *Source*: http://en.wikipedia.org/w/index.php?title=File:37_Fides_symbol.svg *License*: unknown *Contributors*: R. Luther

**Image:243 ida.jpg** *Source*: http://en.wikipedia.org/w/index.php?title=File:243_ida.jpg *License*: unknown *Contributors*: NASA/JPL

**Image:Asteroid 2004 FH.gif** *Source*: http://en.wikipedia.org/w/index.php?title=File:Asteroid_2004_FH.gif *License*: unknown *Contributors*: Images obtained by Stefano Sposetti, Switzerland on March 18, 2004. Animation made Raoul Behrend, Geneva Observatory, Switzerland.

**Image:Moon and Asteroids 1 to 10.svg** *Source*: http://en.wikipedia.org/w/index.php?title=File:Moon_and_Asteroids_1_to_10.svg *License*: unknown *Contributors*: User:Vystrix Nexoth

**Image:Ceres optimized.jpg** *Source*: http://en.wikipedia.org/w/index.php?title=File:Ceres_optimized.jpg *License*: unknown *Contributors*: NASA, ESA, J. Parker (Southwest Research Institute), P. Thomas (Cornell University), and L. McFadden (University of Maryland, College Park)

**Image:masses of asteroids vs main belt.png** *Source*: http://en.wikipedia.org/w/index.php?title=File:Masses_of_asteroids_vs_main_belt.png *License*: unknown *Contributors*: User:Kwamikagami

**Image:(253) mathilde.jpg** *Source*: http://en.wikipedia.org/w/index.php?title=File:(253)_mathilde.jpg *License*: unknown *Contributors*: NASA

**Image:433eros.jpg** *Source*: http://en.wikipedia.org/w/index.php?title=File:433eros.jpg *License*: unknown *Contributors*: NASA/NEAR Project (JHU/APL).

**File:Dawn-image-070911.jpg** *Source*: http://en.wikipedia.org/w/index.php?title=File:Dawn-image-070911.jpg *License*: unknown *Contributors*: NASA/JPL-Caltech/UCLA/MPS/DLR/IDA

**File:Ceres optimized.jpg** *Source*: http://en.wikipedia.org/w/index.php?title=File:Ceres_optimized.jpg *License*: unknown *Contributors*: NASA, ESA, J. Parker (Southwest Research Institute), P. Thomas (Cornell University), and L. McFadden (University of Maryland, College Park)

**File:Pluto-map-hs-2010-06-d270.jpg** *Source*: http://en.wikipedia.org/w/index.php?title=File:Pluto-map-hs-2010-06-d270.jpg *License*: unknown *Contributors*: NASA, ESA, and Marc W. Buie (Southwest Research Institute)

**File:2003EL61art.jpg** *Source*: http://en.wikipedia.org/w/index.php?title=File:2003EL61art.jpg *License*: unknown *Contributors*: A. Feild (Space Telescope Science Institute)

**File:2005FY9art.jpg** *Source*: http://en.wikipedia.org/w/index.php?title=File:2005FY9art.jpg *License*: unknown *Contributors*: Ann Feild (Space Telescope Science Institute)

**File:Eris and dysnomia2.jpg** *Source*: http://en.wikipedia.org/w/index.php?title=File:Eris_and_dysnomia2.jpg *License*: unknown *Contributors*: Ford TTT, Jan.Kamenicek, John800104, Kalki, Krinkle, Laim, Ruslik0, 2 anonymous edits

**File:Masses of dwarf planets vs Luna.png** *Source*: http://en.wikipedia.org/w/index.php?title=File:Masses_of_dwarf_planets_vs_Luna.png *License*: unknown *Contributors*: kwami (talk) Original uploader was Kwamikagami at en.wikipedia Later version(s) were uploaded by Ckatz at en.wikipedia.

**File:Ceres symbol.svg** *Source*: http://en.wikipedia.org/w/index.php?title=File:Ceres_symbol.svg *License*: unknown *Contributors*: Aquantrum, Crei0, Galphi, Lexicon, Ramiy, Ruslik0, Sarang, Sarilho1, Urhixidur, 4 anonymous edits

**File:Pluto symbol.svg** *Source*: http://en.wikipedia.org/w/index.php?title=File:Pluto_symbol.svg *License*: unknown *Contributors*: Lexicon

**File:TheTransneptunians Size Albedo Color.svg** *Source*: http://en.wikipedia.org/w/index.php?title=File:TheTransneptunians_Size_Albedo_Color.svg *License*: unknown *Contributors*: Chesnok, Eurocommuter, Josex.es, Ldanielb00, Poulpy, 9 anonymous edits

**Image:Toutatis.jpg** *Source*: http://en.wikipedia.org/w/index.php?title=File:Toutatis.jpg *License*: unknown *Contributors*: Steve Ostro, JPL

**Image:ESO-Asteroid Toutatis-phot-28c-04-normal.jpg** *Source*: http://en.wikipedia.org/w/index.php?title=File:ESO-Asteroid_Toutatis-phot-28c-04-normal.jpg *License*: unknown *Contributors*: ESO

**File:WISE Finds Fewer Asteroids near Earth.ogv** *Source*: http://en.wikipedia.org/w/index.php?title=File:WISE_Finds_Fewer_Asteroids_near_Earth.ogv *License*: unknown *Contributors*: NASA

**Image:Impact event.jpg** *Source*: http://en.wikipedia.org/w/index.php?title=File:Impact_event.jpg *License*: unknown *Contributors*: Original uploader was Fredrik at en.wikipedia

**File:1950da color 150.jpg** *Source*: http://en.wikipedia.org/w/index.php?title=File:1950da_color_150.jpg *License*: unknown *Contributors*: ComputerHotline, Danim, PedroPVZ, Rmhermen

**Image:Neo-chart.png** *Source*: http://en.wikipedia.org/w/index.php?title=File:Neo-chart.png *License*: unknown *Contributors*: Alan B. Chamberlin

**Image:Outersolarsystem objectpositions labels comp.png** *Source*: http://en.wikipedia.org/w/index.php?title=File:Outersolarsystem_objectpositions_labels_comp.png *License*: unknown *Contributors*: 84user, Kaldari, Peteforsyth, Poppy, Venkat.athma, Wikibob, WilyD, م‌ب‌س‌ا, 7 anonymous edits

**Image:TheKuiperBelt 42AU Centaurs.svg** *Source*: http://en.wikipedia.org/w/index.php?title=File:TheKuiperBelt_42AU_Centaurs.svg *License*: unknown *Contributors*: User:Eurocommuter

**File:AsbolA.GIF** *Source*: http://en.wikipedia.org/w/index.php?title=File:AsbolA.GIF *License*: unknown *Contributors*: Aldo Vitagliano

**Image:TheKuiperBelt Albedo and Color.svg** *Source*: http://en.wikipedia.org/w/index.php?title=File:TheKuiperBelt_Albedo_and_Color.svg *License*: unknown *Contributors*: User:Eurocommuter

**Image:Comet38P2067.gif** *Source*: http://en.wikipedia.org/w/index.php?title=File:Comet38P2067.gif *License*: unknown *Contributors*: Orbit Viewer applet originally written and kindly provided by Osamu Ajiki (AstroArts), and further modified by Ron Baalke (JPL). Tweaks by Kevin Heider.

**File:Lagrange very massive.svg** *Source*: http://en.wikipedia.org/w/index.php?title=File:Lagrange_very_massive.svg *License*: unknown *Contributors*: Antonsusi, EnEdC, JorisvS, Pieter Kuiper, Tano4595, 2 anonymous edits

**File:InnerSolarSystem-en.png** *Source*: http://en.wikipedia.org/w/index.php?title=File:InnerSolarSystem-en.png *License*: unknown *Contributors*: Original uploader was Mdf at en.wikipedia

**File:Solar System size to scale.svg** *Source*: http://en.wikipedia.org/w/index.php?title=File:Solar_System_size_to_scale.svg *License*: unknown *Contributors*: User:Cmglee, User:Farry

**File:Oort cloud Sedna orbit.svg** *Source*: http://en.wikipedia.org/w/index.php?title=File:Oort_cloud_Sedna_orbit.svg *License*: unknown *Contributors*: User:Holek

**Image:Solarsystem3DJupiter.gif** *Source*: http://en.wikipedia.org/w/index.php?title=File:Solarsystem3DJupiter.gif *License*: unknown *Contributors*: User:Lookang

**File:Solarnebula.jpg** *Source*: http://en.wikipedia.org/w/index.php?title=File:Solarnebula.jpg *License*: unknown *Contributors*: William K. Hartmann, Planetary Science Institute, Tucson, Arizona UNIQ-ref-0-9b692a2ad0fef857-QINU

**File:Venustransit 2004-06-08 07-49.jpg** *Source*: http://en.wikipedia.org/w/index.php?title=File:Venustransit_2004-06-08_07-49.jpg *License*: unknown *Contributors*: Ahoerstemeier, Avian, ComputerHotline, Crux, Dbenzhuser, Duckysmokton, Enirac Sum, Infrogmation, J 1982, J.delanoy, KuboF, LA2, Martwinn, Rat at WikiFur, Skab, 1 anonymous edits

**File:Heliospheric-current-sheet.gif** *Source*: http://en.wikipedia.org/w/index.php?title=File:Heliospheric-current-sheet.gif *License*: unknown *Contributors*: Werner Heil (see "other version" below).

**File:4 Terrestrial Planets Size Comp True Color.png** *Source*: http://en.wikipedia.org/w/index.php?title=File:4_Terrestrial_Planets_Size_Comp_True_Color.png *License*: unknown *Contributors*: User:Scooter20

**File:Gas giants in the solar system.jpg** *Source*: http://en.wikipedia.org/w/index.php?title=File:Gas_giants_in_the_solar_system.jpg *License*: unknown *Contributors*: NASA

**File:Comet c1995o1.jpg** *Source*: http://en.wikipedia.org/w/index.php?title=File:Comet_c1995o1.jpg *License*: unknown *Contributors*: User:Miketsukunibito

**File:Outersolarsystem objectpositions labels comp.png** *Source*: http://en.wikipedia.org/w/index.php?title=File:Outersolarsystem_objectpositions_labels_comp.png *License*: unknown *Contributors*: 84user, Kaldari, Peteforsyth, Poppy, Venkat.athma, Wikibob, WilyD, م‌ب‌س‌ا, 7 anonymous edits

**File:EightTNOs.png** *Source*: http://en.wikipedia.org/w/index.php?title=File:EightTNOs.png *License*: unknown *Contributors*: User:Lexicon

**File:IBEX all sky map.jpg** *Source*: http://en.wikipedia.org/w/index.php?title=File:IBEX_all_sky_map.jpg *License*: unknown *Contributors*: NASA/Goddard Space Flight Center Scientific Visualization Studio

**File:Kuiper oort.jpg** *Source*: http://en.wikipedia.org/w/index.php?title=File:Kuiper_oort.jpg *License*: unknown *Contributors*: NASA

**File:Milky Way Spiral Arm.svg** *Source*: http://en.wikipedia.org/w/index.php?title=File:Milky_Way_Spiral_Arm.svg *License*: unknown *Contributors*: User:Surachit

**File:Earth's Location in the Universe (JPEG).jpg** *Source*: http://en.wikipedia.org/w/index.php?title=File:Earth's_Location_in_the_Universe_(JPEG).jpg *License*: unknown *Contributors*: User:Azcolvin429

**File:Magnify-clip.png** *Source*: http://en.wikipedia.org/w/index.php?title=File:Magnify-clip.png *License*: unknown *Contributors*: User:Erasoft24

**File:TheSun.png** *Source*: http://en.wikipedia.org/w/index.php?title=File:TheSun.png *License*: unknown *Contributors*: NASA

**File:Jupiter on 2010-06-07 (captured by the Hubble Space Telescope).jpg** *Source*: http://en.wikipedia.org/w/index.php?title=File:Jupiter_on_2010-06-07_(captured_by_the_Hubble_Space_Telescope).jpg *License*: unknown *Contributors*: NASA, ESA, M.H. Wong (University of Califoria, Berkeley), H.B. Hammel (Space Science Institute, Boulder, Colo.), A.A. Simon-Miller (Goddard Space Flight Center), and the Jupiter Impact Science Team

**File:Saturn closeup.jpg** *Source*: http://en.wikipedia.org/w/index.php?title=File:Saturn_closeup.jpg *License*: unknown *Contributors*: User:BullBoxerBAB

**File:Uranus2.jpg** *Source*: http://en.wikipedia.org/w/index.php?title=File:Uranus2.jpg *License*: unknown *Contributors*: NASA/JPL/Voyager mission Original uploader was Serendipodous at en.wikipedia

**File:Neptune.jpg** *Source*: http://en.wikipedia.org/w/index.php?title=File:Neptune.jpg *License*: unknown *Contributors*: .

**File:The Earth seen from Apollo 17.jpg** *Source*: http://en.wikipedia.org/w/index.php?title=File:The_Earth_seen_from_Apollo_17.jpg *License*: unknown *Contributors*: NASA/Apollo 17 crew; taken by either Harrison Schmitt or Ron Evans

**File:Venus-real.jpg** *Source*: http://en.wikipedia.org/w/index.php?title=File:Venus-real.jpg *License*: unknown *Contributors*: NASA/Ricardo Nunes

**File:Mars Hubble.jpg** *Source*: http://en.wikipedia.org/w/index.php?title=File:Mars_Hubble.jpg *License*: unknown *Contributors*: NASA and The Hubble Heritage Team (STScI/AURA)

**File:Ganymede g1 true 2.jpg** *Source*: http://en.wikipedia.org/w/index.php?title=File:Ganymede_g1_true_2.jpg *License*: unknown *Contributors*: Original uploader was GeneralPatton at en.wikipedia

**File:Two Halves of Titan.png** *Source*: http://en.wikipedia.org/w/index.php?title=File:Two_Halves_of_Titan.png *License*: unknown *Contributors*: NASA/JPL/Space Science Institute

**File:Mercury in color - Prockter07 centered.jpg** *Source*: http://en.wikipedia.org/w/index.php?title=File:Mercury_in_color_-_Prockter07_centered.jpg *License*: unknown *Contributors*: User:Papa Lima Whiskey

**File:Callisto.jpg** *Source*: http://en.wikipedia.org/w/index.php?title=File:Callisto.jpg *License*: unknown *Contributors*: Bricktop, Conscious, Dbenbenn, J 1982, JenVan, Kristaga, Li-sung, Ruslik0, 1 anonymous edits

**File:Io highest resolution true color.jpg** *Source*: http://en.wikipedia.org/w/index.php?title=File:Io_highest_resolution_true_color.jpg *License*: unknown *Contributors*: Bricktop, David Levy, Foroa, Jed, Julia W, Kriplozoik, Pfctdayelise, Ruslik0, Uwe W., WikipediaMaster, 1 anonymous edits

**File:FullMoon2010.jpg** *Source*: http://en.wikipedia.org/w/index.php?title=File:FullMoon2010.jpg *License*: unknown *Contributors*: Gregory H. Revera

**File:Europa-moon.jpg** *Source*: http://en.wikipedia.org/w/index.php?title=File:Europa-moon.jpg *License*: unknown *Contributors*: Dbenbenn, Li-sung, Ruslik0, Sobi3ch, Wikiborg4711, 2 anonymous edits

**File:Triton Voyager 2.jpg** *Source*: http://en.wikipedia.org/w/index.php?title=File:Triton_Voyager_2.jpg *License*: unknown *Contributors*: Rmhermen, Ruslik0

**File:Titania (moon) color cropped.jpg** *Source*: http://en.wikipedia.org/w/index.php?title=File:Titania_(moon)_color_cropped.jpg *License*: unknown *Contributors*: User:Ruslik0

**File:PIA07763 Rhea full globe5.jpg** *Source*: http://en.wikipedia.org/w/index.php?title=File:PIA07763_Rhea_full_globe5.jpg *License*: unknown *Contributors*: NASA/JPL/Space Science Institute

**File:Voyager 2 picture of Oberon.jpg** *Source*: http://en.wikipedia.org/w/index.php?title=File:Voyager_2_picture_of_Oberon.jpg *License*: unknown *Contributors*: NASA Original uploader was Looxix at en.wikipedia

**File:Iapetus as seen by the Cassini probe - 20071008.jpg** *Source*: http://en.wikipedia.org/w/index.php?title=File:Iapetus_as_seen_by_the_Cassini_probe_-_20071008.jpg *License*: unknown *Contributors*: NASA/JPL/Space Science Institute

**File:Umbriel moon 1.gif** *Source*: http://en.wikipedia.org/w/index.php?title=File:Umbriel_moon_1.gif *License*: unknown *Contributors*: NASA (Voyager)

**File:Ariel-NASA.jpg** *Source*: http://en.wikipedia.org/w/index.php?title=File:Ariel-NASA.jpg *License*: unknown *Contributors*: Original uploader was Svdmolen at nl.wikipedia

**File:Dione (Mond) (30823363).jpg** *Source*: http://en.wikipedia.org/w/index.php?title=File:Dione_(Mond)_(30823363).jpg *License*: unknown *Contributors*: alan taylor from Hudson, MA, usa

**File:Inset-sat tethys-large.jpg** *Source*: http://en.wikipedia.org/w/index.php?title=File:Inset-sat_tethys-large.jpg *License*: unknown *Contributors*: NASA / JPL

**File:Vesta_full_mosaic.jpg** *Source*: http://en.wikipedia.org/w/index.php?title=File:Vesta_full_mosaic.jpg *License*: unknown *Contributors*: JorisvS, Szczureq

**File:Enceladus from Voyager.jpg** *Source*: http://en.wikipedia.org/w/index.php?title=File:Enceladus_from_Voyager.jpg *License*: unknown *Contributors*: NASA/JPL/USGS

**File:Miranda.jpg** *Source*: http://en.wikipedia.org/w/index.php?title=File:Miranda.jpg *License*: unknown *Contributors*: From de.wiki (NASA image)

**File:Proteus Voyager 2 croped.jpg** *Source*: http://en.wikipedia.org/w/index.php?title=File:Proteus_Voyager_2_croped.jpg *License*: unknown *Contributors*: Voyager 2, NASA

**File:Mimas moon.jpg** *Source*: http://en.wikipedia.org/w/index.php?title=File:Mimas_moon.jpg *License*: unknown *Contributors*: NASA

**File:Hyperion in natural colours.jpg** *Source*: http://en.wikipedia.org/w/index.php?title=File:Hyperion_in_natural_colours.jpg *License*: unknown *Contributors*: Uwe W., 1 anonymous edits

**File:Phoebe cassini.jpg** *Source*: http://en.wikipedia.org/w/index.php?title=File:Phoebe_cassini.jpg *License*: unknown *Contributors*: NASA/JPL/Space Science Institute

**File:PIA12714 Janus crop.jpg** *Source*: http://en.wikipedia.org/w/index.php?title=File:PIA12714_Janus_crop.jpg *License*: unknown *Contributors*: NASA / Jet Propulsion Laboratory / Space Science Institute

**File:Amalthea (moon).gif** *Source*: http://en.wikipedia.org/w/index.php?title=File:Amalthea_(moon).gif *License*: unknown *Contributors*: Giorgiomonteforti, Nolanus, Quadell, Ruslik0, Uwe W.

**File:PIA09813 Epimetheus S. polar region.jpg** *Source*: http://en.wikipedia.org/w/index.php?title=File:PIA09813_Epimetheus_S._polar_region.jpg *License*: unknown *Contributors*: NASA/JPL/Space Science Institute

**File:Prometheus 12-26-09a.jpg** *Source*: http://en.wikipedia.org/w/index.php?title=File:Prometheus_12-26-09a.jpg *License*: unknown *Contributors*: NASA / JPL / Space Science Institute

**File:GerardKuiper.jpg** *Source*: http://en.wikipedia.org/w/index.php?title=File:GerardKuiper.jpg *License*: unknown *Contributors*: Dodo, Miraceti, Ruslik0, Svdmolen

**File:Maunatele.jpg** *Source*: http://en.wikipedia.org/w/index.php?title=File:Maunatele.jpg *License*: unknown *Contributors*: NASA

**File:Lhborbits.png** *Source*: http://en.wikipedia.org/w/index.php?title=File:Lhborbits.png *License*: unknown *Contributors*: Jan.Kamenicek, JorisvS, MichaelSchoenitzer, Poppy, Rathorius, Stanlekub, 3 anonymous edits

**File:Dust Models Paint Alien's View of Solar System.ogv** *Source*: http://en.wikipedia.org/w/index.php?title=File:Dust_Models_Paint_Alien's_View_of_Solar_System.ogv *License*: unknown *Contributors*: NASA

**File:TheKuiperBelt 75AU All.svg** *Source*: http://en.wikipedia.org/w/index.php?title=File:TheKuiperBelt_75AU_All.svg *License*: unknown *Contributors*: User:Eurocommuter

**File:TheKuiperBelt classes-en.svg** *Source*: http://en.wikipedia.org/w/index.php?title=File:TheKuiperBelt_classes-en.svg *License*: unknown *Contributors*: User:Eurocommuter, User:Lilyu

**File:ExampleUpdatedHistogramOfTNOsemimajoraxii.png** *Source*: http://en.wikipedia.org/w/index.php?title=File:ExampleUpdatedHistogramOfTNOsemimajoraxii.png *License*: unknown *Contributors*: User:WilyD

**File:2003 UB313 near-infrared spectrum.gif** *Source*: http://en.wikipedia.org/w/index.php?title=File:2003_UB313_near-infrared_spectrum.gif *License*: unknown *Contributors*: Admrboltz, Adoniscik, Bryan Derksen, Chromosome, EDUCA33E, Ewen, Jan.Kamenicek, Orionist, Poppy, Poulpy, مانفي, 1 anonymous edits

**File:TheKuiperBelt PowerLaw2.svg** *Source*: http://en.wikipedia.org/w/index.php?title=File:TheKuiperBelt_PowerLaw2.svg *License*: unknown *Contributors*: User:Eurocommuter

**File:TheKuiperBelt Projections 100AU Classical SDO.svg** *Source*: http://en.wikipedia.org/w/index.php?title=File:TheKuiperBelt_Projections_100AU_Classical_SDO.svg *License*: unknown *Contributors*: User:Eurocommuter

**File:Triton moon mosaic Voyager 2 (large).jpg** *Source*: http://en.wikipedia.org/w/index.php?title=File:Triton_moon_mosaic_Voyager_2_(large).jpg *License*: unknown *Contributors*: NASA / Jet Propulsion Lab / U.S. Geological Survey

**File:New horizons Pluto.jpg** *Source*: http://en.wikipedia.org/w/index.php?title=File:New_horizons_Pluto.jpg *License*: unknown *Contributors*: Bryan Derksen, Dream out loud, Kukanotas, Ruslik0, SchuminWeb, ShakataGaNai, 2 anonymous edits

**File:Kuiper belt remote.jpg** *Source*: http://en.wikipedia.org/w/index.php?title=File:Kuiper_belt_remote.jpg *License*: unknown *Contributors*: User:Audriusa

**GNU Free Documentation License Version 1.2, November 2002 Copyright (C) 2000,2001,2002 Free Software Foundation, Inc. 59 Temple Place, Suite 330, Boston, MA 02111-1307 USA Everyone is permitted to copy and distribute verbatim copies of this license document, but changing it is not allowed.**

**0. PREAMBLE**

The purpose of this License is to make a manual, textbook, or other functional and useful document "free" in the sense of freedom: to assure everyone the effective freedom to copy and redistribute it, with or without modifying it, either commercially or noncommercially. Secondarily, this License preserves for the author and publisher a way to get credit for their work, while not being considered responsible for modifications made by others. This License is a kind of "copyleft", which means that derivative works of the document must themselves be free in the same sense. It complements the GNU General Public License, which is a copyleft license designed for free software. We have designed this License in order to use it for manuals for free software, because free software needs free documentation: a free program should come with manuals providing the same freedoms that the software does. But this License is not limited to software manuals; it can be used for any textual work, regardless of subject matter or whether it is published as a printed book. We recommend this License principally for works whose purpose is instruction or reference.

**1. APPLICABILITY AND DEFINITIONS**

This License applies to any manual or other work, in any medium, that contains a notice placed by the copyright holder saying it can be distributed under the terms of this License. Such a notice grants a world-wide, royalty-free license, unlimited in duration, to use that work under the conditions stated herein. The "Document", below, refers to any such manual or work. Any member of the public is a licensee, and is addressed as "you". You accept the license if you copy, modify or distribute the work in a way requiring permission under copyright law. A "Modified Version" of the Document means any work containing the Document or a portion of it, either copied verbatim, or with modifications and/or translated into another language. A "Secondary Section" is a named appendix or a front-matter section of the Document that deals exclusively with the relationship of the publishers or authors of the Document to the Document's overall subject (or to related matters) and contains nothing that could fall directly within that overall subject. (Thus, if the Document is in part a textbook of mathematics, a Secondary Section may not explain any mathematics.) The relationship could be a matter of historical connection with the subject or with related matters, or of legal, commercial, philosophical, ethical or political position regarding them. The "Invariant Sections" are certain Secondary Sections whose titles are designated, as being those of Invariant Sections, in the notice that says that the Document is released under this License. If a section does not fit the above definition of Secondary then it is not allowed to be designated as Invariant. The Document may contain zero Invariant Sections. If the Document does not identify any Invariant Sections then there are none. The "Cover Texts" are certain short passages of text that are listed, as Front-Cover Texts or Back-Cover Texts, in the notice that says that the Document is released under this License. A Front-Cover Text may be at most 5 words, and a Back-Cover Text may be at most 25 words. A "Transparent" copy of the Document means a machine-readable copy, represented in a format whose specification is available to the general public, that is suitable for revising the document straightforwardly with generic text editors or (for images composed of pixels) generic paint programs or (for drawings) some widely available drawing editor, and that is suitable for input to text formatters or for automatic translation to a variety of formats suitable for input to text formatters. A copy made in an otherwise Transparent file format whose markup, or absence of markup, has been arranged to thwart or discourage subsequent modification by readers is not Transparent. An image format is not Transparent if used for any substantial amount of text. A copy that is not "Transparent" is called "Opaque". Examples of suitable formats for Transparent copies include plain ASCII without markup, Texinfo input format, LaTeX input format, SGML or XML using a publicly available DTD, and standard-conforming simple HTML, PostScript or PDF designed for human modification. Examples of transparent image formats include PNG, XCF and JPG. Opaque formats include proprietary formats that can be read and edited only by proprietary word processors, SGML or XML for which the DTD and/or processing tools are not generally available, and the machine-generated HTML, PostScript or PDF produced by some word processors for output purposes only. The "Title Page" means, for a printed book, the title page itself, plus such following pages as are needed to hold, legibly, the material this License requires to appear in the title page. For works in formats which do not have any title page as such, "Title Page" means the text near the most prominent appearance of the work's title, preceding the beginning of the body of the text. A section "Entitled XYZ" means a named subunit of the Document whose title either is precisely XYZ or contains XYZ in parentheses following text that translates XYZ in another language. (Here XYZ stands for a specific section name mentioned below, such as "Acknowledgements", "Dedications", "Endorsements", or "History".) To "Preserve the Title" of such a section when you modify the Document means that it remains a section "Entitled XYZ" according to this definition. The Document may include Warranty Disclaimers next to the notice which states that this License applies to the Document. These Warranty Disclaimers are considered to be included by reference in this License, but only as regards disclaiming warranties: any other implication that these Warranty Disclaimers may have is void and has no effect on the meaning of this License.

**2. VERBATIM COPYING**

You may copy and distribute the Document in any medium, either commercially or noncommercially, provided that this License, the copyright notices, and the license notice saying this License applies to the Document are reproduced in all copies, and that you add no other conditions whatsoever to those of this License. You may not use technical measures to obstruct or control the reading or further copying of the copies you make or distribute. However, you may accept compensation in exchange for copies. If you distribute a large enough number of copies you must also follow the conditions in section 3. You may also lend copies, under the same conditions stated above, and you may publicly display copies.

**3. COPYING IN QUANTITY**

If you publish printed copies (or copies in media that commonly have printed covers) of the Document, numbering more than 100, and the Document's license notice requires Cover Texts, you must enclose the copies in covers that carry, clearly and legibly, all these Cover Texts: Front-Cover Texts on the front cover, and Back-Cover Texts on the back cover. Both covers must also clearly and legibly identify you as the publisher of these copies. The front cover must present the full title with all words of the title equally prominent and visible. You may add other material on the covers in addition. Copying with changes limited to the covers, as long as they preserve the title of the Document and satisfy these conditions, can be treated as verbatim copying in other respects. If the required texts for either cover are too voluminous to fit legibly, you should put the first ones listed (as many as fit reasonably) on the actual cover, and continue the rest onto adjacent pages. If you publish or distribute Opaque copies of the Document numbering more than 100, you must either include a machine-readable Transparent copy along with each Opaque copy, or state in or with each Opaque copy a computer-network location from which the general network-using public has access to download using public-standard network protocols a complete Transparent copy of the Document, free of added material. If you use the latter option, you must take reasonably prudent steps, when you begin distribution of Opaque copies in quantity, to ensure that this Transparent copy will remain thus accessible at the stated location until at least one year after the last time you distribute an Opaque copy (directly or through your agents or retailers) of that edition to the public. It is requested, but not required, that you contact the authors of the Document well before redistributing any large number of copies, to give them a chance to provide you with an updated version of the Document.

**4. MODIFICATIONS**

You may copy and distribute a Modified Version of the Document under the conditions of sections 2 and 3 above, provided that you release the Modified Version under precisely this License, with the Modified Version filling the role of the Document, thus licensing distribution and modification of the Modified Version to whoever possesses a copy of it. In addition, you must do these things in the Modified Version: A. Use in the Title Page (and on the covers, if any) a title distinct from that of the Document, and from those of previous versions (which should, if there were any, be listed in the History section of the Document). You may use the same title as a previous version if the original publisher of that version gives permission. B. List on the Title Page, as authors, one or more persons or entities responsible for authorship of the modifications in the Modified Version, together with at least five of the principal authors of the Document (all of its principal authors, if it has fewer than five), unless they release you from this requirement. C. State on the Title page the name of the publisher of the Modified Version, as the publisher. D. Preserve all the copyright notices of the Document. E. Add an appropriate copyright notice for your modifications adjacent to the other copyright notices. F. Include, immediately after the copyright notices, a license notice giving the public permission to use the Modified Version under the terms of this License, in the form shown in the Addendum below. G. Preserve in that license notice the full lists of Invariant Sections and required Cover Texts given in the Document's license notice. H. Include an unaltered copy of this License. I. Preserve the section Entitled "History", Preserve its Title, and add to it an item stating at least the title, year, new authors, and publisher of the Modified Version as given on the Title Page. If there is no section Entitled "History" in the Document, create one stating the title, year, authors, and publisher of the Document as given on its Title Page, then add an item describing the Modified Version as stated in the previous sentence. J. Preserve the network location, if any, given in the Document for public access to a Transparent copy of the Document, and likewise the network locations given in the Document for previous versions it was based on. These may be placed in the "History" section. You may omit a network location for a work that was published at least four years before the Document itself, or if the original publisher of the version it refers to gives permission. K. For any section Entitled "Acknowledgements" or "Dedications", Preserve the Title of the section, and preserve in the section all the substance and tone of each of the contributor acknowledgements and/or dedications given therein. L. Preserve all the Invariant Sections of the Document, unaltered in their text and in their titles. Section numbers or the equivalent are not considered part of the section titles. M. Delete any section Entitled "Endorsements". Such a section may not be included in the Modified Version. N. Do not retitle any existing section to be Entitled "Endorsements" or to conflict in title with any Invariant Section. O. Preserve any Warranty Disclaimers. If the Modified Version includes new front-matter sections or appendices that qualify as Secondary Sections and contain no material copied from the Document, you may at your option designate some or all of these sections as invariant. To do this, add their titles to the list of Invariant Sections in the Modified Version's license notice. These titles must be distinct from any other section titles. You may add a section Entitled "Endorsements", provided it contains nothing but endorsements of your Modified Version by various parties--for example, statements of peer review or that the text has been approved by an organization as the authoritative definition of a standard. You may add a passage of up to five words as a Front-Cover Text, and a passage of up to 25 words as a Back-Cover Text, to the end of the list of Cover Texts in the Modified Version. Only one passage of Front-Cover Text and one of Back-Cover Text may be added by (or through arrangements made by) any one entity. If the Document already includes a cover text for the same cover, previously added by you or by arrangement made by the same entity you are acting on behalf of, you may not add another; but you may replace the old one, on explicit permission from the previous publisher that added the old one. The author(s) and publisher(s) of the Document do not by this License give permission to use their names for publicity for or to assert or imply endorsement of any Modified Version.

**5. COMBINING DOCUMENTS**

You may combine the Document with other documents released under this License, under the terms defined in section 4 above for modified versions, provided that you include in the combination all of the Invariant Sections of all of the original documents, unmodified, and list them all as Invariant Sections of your combined work in its license notice, and that you preserve all their Warranty Disclaimers. The combined work need only contain one copy of this License, and multiple identical Invariant Sections may be replaced with a single copy. If there are multiple Invariant Sections with the same name but different contents, make the title of each such section unique by adding at the end of it, in parentheses, the name of the original author or publisher of that section if known, or else a unique number. Make the same adjustment to the section titles in the list of Invariant Sections in the license notice of the combined work. In the combination, you must combine any sections Entitled "History" in the various original documents, forming one section Entitled "History"; likewise combine any sections Entitled "Acknowledgements", and any sections Entitled "Dedications". You must delete all sections Entitled "Endorsements".

**6. COLLECTIONS OF DOCUMENTS**

You may make a collection consisting of the Document and other documents released under this License, and replace the individual copies of this License in the various documents with a single copy that is included in the collection, provided that you follow the rules of this License for verbatim copying of each of the documents in all other respects. You may extract a single document from such a collection, and distribute it individually under this License, provided you insert a copy of this License into the extracted document, and follow this License in all other respects regarding verbatim copying of that document.

**7. AGGREGATION WITH INDEPENDENT WORKS**

A compilation of the Document or its derivatives with other separate and independent documents or works, in or on a volume of a storage or distribution medium, is called an "aggregate" if the copyright resulting from the compilation is not used to limit the legal rights of the compilation's users beyond what the individual works permit. When the Document is included in an aggregate, this License does not apply to the other works in the aggregate which are not themselves derivative works of the Document. If the Cover Text requirement of section 3 is applicable to these copies of the Document, then if the Document is less than one half of the entire aggregate, the Document's Cover Texts may be placed on covers that bracket the Document within the aggregate, or the electronic equivalent of covers if the Document is in electronic form. Otherwise they must appear on printed covers that bracket the whole aggregate.

**8. TRANSLATION**

Translation is considered a kind of modification, so you may distribute translations of the Document under the terms of section 4. Replacing Invariant Sections with translations requires special permission from their copyright holders, but you may include translations of some or all Invariant Sections in addition to the original versions of these Invariant Sections. You may include a translation of this License, and all the license notices in the Document, and any Warranty Disclaimers, provided that you also include the original English version of this License and the original versions of those notices and disclaimers. In case of a disagreement between the translation and the original version of this License or a notice or disclaimer, the original version will prevail. If a section in the Document is Entitled "Acknowledgements", "Dedications", or "History", the requirement (section 4) to Preserve its Title (section 1) will typically require changing the actual title.

**9. TERMINATION**

You may not copy, modify, sublicense, or distribute the Document except as expressly provided for under this License. Any other attempt to copy, modify, sublicense or distribute the Document is void, and will automatically terminate your rights under this License. However, parties who have received copies, or rights, from you under this License will not have their licenses terminated so long as such parties remain in full compliance.

**10. FUTURE REVISIONS OF THIS LICENSE**

The Free Software Foundation may publish new, revised versions of the GNU Free Documentation License from time to time. Such new versions will be similar in spirit to the present version, but may differ in detail to address new problems or concerns. See http://www.gnu.org/copyleft/. Each version of the License is given a distinguishing version number. If the Document specifies that a particular numbered version of this License "or any later version" applies to it, you have the option of following the terms and conditions either of that specified version or of any later version that has been published (not as a draft) by the Free Software Foundation. If the Document does not specify a version number of this License, you may choose any version ever published (not as a draft) by the Free Software Foundation. ADDENDUM: How to use this License for your documents To use this License in a document you have written, include a copy of the License in the document and put the following copyright and license notices just after the title page: Copyright (c) YEAR YOUR NAME. Permission is granted to copy, distribute and/or modify this document under the terms of the GNU Free Documentation License, Version 1.2 or any later version published by the Free Software Foundation; with no Invariant Sections, no Front-Cover Texts, and no Back-Cover Texts. A copy of the license is included in the section entitled "GNU Free Documentation License". If you have Invariant Sections, Front-Cover Texts and Back-Cover Texts, replace the "with...Texts." line with this: with the Invariant Sections being LIST THEIR TITLES, with the Front-Cover Texts being LIST, and with the Back-Cover Texts being LIST. If you have Invariant Sections without Cover Texts, or some other combination of the three, merge those two alternatives to suit the situation. If your document contains nontrivial examples of program code, we recommend releasing these examples in parallel under your choice of free software license, such as the GNU General Public License, to permit their use in free software.

MIX
Papier aus verantwortungsvollen Quellen
Paper from responsible sources
FSC® C105338

Printed by Books on Demand GmbH, Norderstedt / Germany